MODERN FORMAL METHODS AND APPLICATIONS

Modern Formal Methods and Applications

Edited by

HOSSAM A. GABBAR
Okayama University, Okayama, Japan

 Springer

A C.I.P. Catalogue record for this book is available from the Library of Congress.

ISBN-10 1-4020-4222-1 (HB)
ISBN-13 978-1-4020-4222-5 (HB)
ISBN-10 1-4020-4223-X (e-book)
ISBN-13 978-1-4020-4223-2 (e-book)

Published by Springer,
P.O. Box 17, 3300 AA Dordrecht, The Netherlands.

www.springer.com

Printed on acid-free paper

Printed in the Netherlands.

Table of Contents

Preface

Problem solving techniques and methods are widely used in different disciplines such as medical, engineering, social, marine, business, etc., where it is required to find optimum or desired solution in the problem domain. Formal methods are introduced as a way to transform the problem from the informal space to the formal space where it becomes easier for computational methods and technologies to be adopted and applied to solve the underlying problem.

Formal methods, or automata theory, were originally introduced in the 1930's by computer scientists, mathematicians, and linguistics. The area of formal methods includes subjects such as: formal languages, formal specifications, predicate logic, knowledge representation, and automata. In principle, formal methods are used to describe the problem in a way that will help in finding the solution. Initially, it is widely used with software engineering to specify the target system to be able to design, develop, and to validate the underlying system. Formal methods are also used with product/process design, and with biological and human diagnostic applications.

Studying formal methods as a pure science is necessary but not adequate to address the relevant problems and to get hands on in the different disciplines. To address the variety of needs, this book provides basic concepts of formal methods and presents state-of-the art methods and their applications to critical problems in different disciplines such as engineering, natural resource & waste management, production chain management, biological systems, software, and hardware related problems.

This book will give both high-level and insights for specialists and practitioners who are interested in the area of formal methods and willing to apply these methods in their problem domains.

This book is organized in chapters that will give the reader basic background about formal languages, in Chapter 1. Chapters 2 and 3 show some practical applications of formal methods in engineering and supply chain. Chapter 4 describes another important engineering application of waste management practices. And in completely different disciplines, forma methods are using for biological network modeling (Chapter 5), software specifications (Chapter 6), and for hardware compilation (Chapter 7). The final chapter (Chapter 8) of this book shows another example of the use of formal language to formally specify UML for modeling and validating system models.

Acknowledgement

It is our pleasure to produce such useful scientific communication in the area of formal methods, which we believe will have great value to all categories of readers from undergraduate and researchers as well as from industry who can apply theses methods, practices, and examples to their own cases.

About the Editor

Hossam A. Gabbar is an Associate Professor in the Graduate School of Natural Science & Technology, Division of Industrial Innovative Sciences, Okayama University, Japan. In 1988, he obtained his B.Sc. from Computer Science & Automatic Control, Faculty of Engineering, Alexandria University (Egypt), with final grade distinction with class of honor. In 1990, he completed the master courses from the same department. He worked as a software engineer, IT project manager, and senior consultant in several industrial projects in different disciplines such as: oil & gas, manufacturing, investment, telecomm, marine, and chemical/pharmaceutical industries. In the academic side, he worked in research centers in the areas of marine supply automation and coast protection. He joined Tokyo Institute of Technology and Japan Chemical Innovative Institute (JCII), where he participated in national projects related to batch plant control, oil & gas offsite systems, biomass production systems, and plastic production chain with recycling. His research focus and interests include process systems engineering, where he investigates learning systems for process, safety and risk management, plant operation, maintenance management, and fault diagnostic with the considerations of life cycle activities. His recent research interests include integrated resource planning system, and its application on the planning of future energy production systems.

He is elected as senior member of IEEE in 2004 for his research achievements and contribution to industry. He is a member in several scientific organizations such as IEEE-CSS / SMC, Japan Ergonomics Society (JES), Society of Chemical Engineering – Japan (SCEJ), and Society of Instrument and Control Engineers (SICE). He is the author/co-author of more than 50 recognized publications, including papers, books, and technical reports. He is the first-name inventor of a patent for batch recipe synthesis and control, which has been implemented successfully in Mitsubishi Chemicals Co., Japan, and acquired by Japanese venture company to be realized in Japanese and international industries. His recent research achievements are the development of innovative solution for oil & gas offsite process control and operation, with Yokogawa Co., Japan.

About the Authors

Adrien Richard
Adrien Richard was born in January 1968, France. In 2003, he obtained his Ph.D. degree in D.E.A. Mathematics and Computer Science, applied to Biology from CNRS and University of Evry, France. Since 2002, he was engaged in research in modeling of biological regulatory networks and in the verification of properties of such systems.

Carlos E. Cuesta
Carlos E. Cuesta received the B.Sc., M.Eng. and Ph.D. degrees in Computer Science from the University of Valladolid, Spain. He is presently an Associate Professor in the Department of Computer Science of the University of Valladolid, and his research interests are software architecture and engineering, and formal methods.

He Jifeng
Professor He Jifeng has been a Senior Research Fellow at the United Nations University International Institute for Software Technology (UNU/IIST) in Macau since 1998. Between 1984 and 1998 he was a senior researcher at the Oxford University Computing Laboratory Programming Research Group in England where he worked extensively with Sir Tony Hoare. His research interest lies in the sound methods of specification of computer systems, communications, application and standards, and the techniques for designing and implementing those specifications in software and/or hardware, with high reliability and at low cost. He has authored books on "Provably Correct Systems" and (with Tony Hoare) "Unifying Theories of Programming" as well as numerous research papers. He is Professor of Computer Science at two Chinese universities, East China Normal University since 1986 and Shanghai

Jiao Tong University since 1996.

Jean-Paul Comet

Professor Jean-Paul Comet was born in October 1968, France. In 1992, he has graduated in Applied Computer Science Engineering from the Department of applied mathematics of INSA-Rouen, France. He was awarded the Ph.D. degree in 1998 in Computer Science at The French National Institute for Research in Computer Science and Control (INRIA). He spent 1 year at The Whitehead Institute/MIT Center for Genome Research (Cambridge (MA), USA) in 1999. Jean-Paul Comet is now Assistant Professor at the University of Evry-Val d'Essonne, France. During 1994-1998, he was engaged in research in the field of biological sequence comparison, where he participated in research program for DNA chips data analysis. He is now interested in the modeling of biological regulatory networks and in the verification of properties of such systems.

Jonathan Bowen

Professor Jonathan Bowen is at London South Bank University, UK, where he is Professor of Computing and heads the Centre for Applied Formal Methods. From 1995 to March 2000, Bowen was a lecturer at the Department of Computer Science, University of Reading where he led the Formal Methods and Software Engineering Group. Previously he was a senior researcher at the Oxford University Computing Laboratory Programming Research Group where he worked under the guidance of Sir Tony Hoare. Between 1979 and 1984 he worked at Imperial College, London as a research assistant, latterly in the interdepartmental Wolfson Microprocessor Laboratory. He has been involved with the fields of electronics and computing in both industry (including Marconi Instruments, Logica and Silicon Graphics Inc.) and academia since 1977. His interests include formal methods, safety-critical systems, the Z notation, provably correct systems, rapid prototyping using logic programming, decompilation, hardware compilation, software/hardware co-design, the history of computing and online museums. He has produced around 250 publications, including 13 books, and has served on over 60 programme committees. During 2001 Bowen received the Freedom of the Worshipful Company of Information Technologists. In 2002, Bowen was elected Chair of the British Computer Society FACS Specialist Group on Formal Aspects of Computing Science and Fellow of the Royal Society for the Arts. He is a member of the ACM and IEEE Computer Society, and holds an MA degree in Engineering Science from Oxford University.

M. Encarnacion Beato

M. Encarnacion Beato received the M.Eng. and Ph.D. degrees in computer science from the University of Valladolid, Spain. He is presently Associate

Professor in the Department of Computer Science of the University Pontificia of Salamanca, and his research interests are Software Engineering and formal methods.

Manuel Barrio-Solorzano

Manuel Barrio-Sol zano received the Ph.D. degree in computer science from the University of Valladolid, Spain. He is presently Associate Professor in the Department of Computer Science of the University of Valladolid, and his research interests are software engineering and formal methods.

Pablo de la Fuente

Pablo de la Fuente received the Ph.D. degree in 1989 from the University of Valladolid, Spain. He is an associate professor in the Computer Science Department at the University Valladolid, Spain. His main research interests are in the areas of architecture-based software development and text retrieval. He is a member of the ACM, and IEEE Computer Society.

Veikko Pohjola

Professor Veikko Pohjola, obtained his M.Sc. (Eng) from Helsinki University of Technology in 1967. In 1970, he obtained the Lic.Tech from Helsinki University of Technology. And in 1974, he obtained the D.Tech from Helsinki University of Technology. From 1986 till 2004, he was Pofessor of Chemical Process Engineering at the University of Oulu. He is a member of the Board of Nordem Oy since 2001. Currently, he is Emeritus professor at the University of Oulu since 2004. His recent research interests include the development and applications of PSSP ontology based methods in various fields including, knowledge management, education and training, and Design.

List of Figures

List of Tables

1 Fundamentals of Formal Methods

Author
Hossam A.Gabbar,
Graduate School of Natural Science & Technology,
Okayama University

Summary
Formal methods, specifications, and languages are used as part of the problem-solving paradigm. This chapter presents basics of these subjects and presenting the progress and advances made until now.

Keywords: formal methods, learning, reasoning, formal specifications, formal languages.

1.1 Overview

Sometimes it is illuminating to go back to the origin of a word and this is indeed the case: "method" comes from Greek and means "way through"; the Latin substitute for it quite significantly is "via et ratio" but also "ratio et via", both conveying the meaning of "something rational with the purpose of achieving something, together with the way of achieving it".

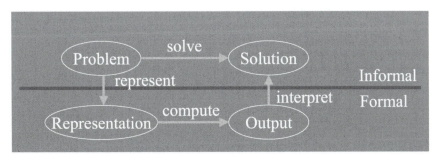

Figure 1-1. Problem Solving Framework

Formalism doesn't mean method; in principle formalism can be associated with different methods, or lead to no useful method at all. The combined terms

1

Hossam A. Gabbar (ed.), Modern Formal Methods and Applications, 1–20.
© 2006 *Springer. Printed in the Netherlands.*

'Formal Methods' is regarded as the use of formality to represent methods that are used in/for process or system engineering. Formal methods are practical and precise way of solving problems [4]. Figure 1-1 shows basic problem solving framework, which comprises formal and informal domains. It is quite important to find suitable and comprehensive way to define and describe the underlying problem so that it becomes easier to find solution.

The formal methods can be viewed as the formal way to describe problem or to model system. One of the first definitions of formal methods is: A broad view of formal methods includes all applications of (primarily) discrete mathematics to software engineering problems. This application usually involves modeling and analysis where the models and analysis procedures are derived from or defined by an underlying mathematically- precise foundation. [Leveson 90]. Such definition is limited to software problems, however, it has been dramatically extended to include biological systems, engineering systems, social systems, and other disciplines. Formal methods support precise and rigorous specifications of those aspects of the underlying system capable of being used to manage the system throughout its life cycle.

This can be extended to another advanced definition of formal methods as: mathematically based techniques for the specification, development and verification of the underlying system.

Formal methods can include graphical languages. For example, Data Flow Diagrams (DFDs) are the most well-known graphical technique for specifying the function of a system. DFDs can be considered a semi-formal method, and researchers have explored techniques for treating DFDs in a completely formal manner. Petri nets provide another well-known graphical technique, often used in distributed systems [Peterson 77]. Another example is Petri nets, which are a fully formal technique. Yet, another formal method is the Finite state machines, which are commonly presented in tabular form.

There is an increasing interest about formal methods and their applications, where mathematical synthesis and analysis techniques are applied to the development of (computer) controlled systems. Although formal methods were around for more than 25 years, but their use was limited to software and hardware systems. The recent years witnessed revolutionary computational systems where intelligent systems are required to manage complex systems in different industrial disciplines. Formal methods have the potential to provide increased confidence in a system by satisfying the standards set by regulatory bodies.

Formality level can be varied from application to application and from domain to domain, based on the requirements and available specification detailed level. Figure 1-2 shows different levels of formalization spectrum. In such figure specification language is used as a set of formulae in a formal language to describe the underlying system.

Objects that satisfy a given specification in the semantic domain of a given language can be non-unique. Several objects may be equivalent as far as a particular specification is concerned. Because of this non-uniqueness, the specification is at a higher level of abstraction than the objects in the semantic domain. The specification language permits abstraction from details that distinguish different implementations, while preserving essential properties.

Different specification methods defined over the same semantic domain allow for specifying different aspects of specified objects. These concepts can be defined more precisely using mathematics. The advantage of this mathematics is that it provides tools for formal reasoning about specifications. Specifications can then be examined for completeness and consistency.

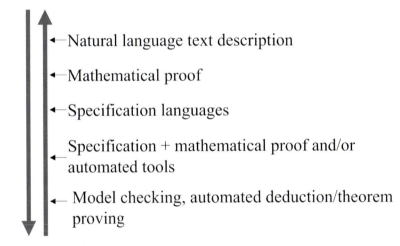

Less Formal

←Natural language text description

←Mathematical proof

←Specification languages

←Specification + mathematical proof and/or automated tools

←Model checking, automated deduction/theorem proving

More formal

Figure 1-2. Formalization Spectrum

Specification languages can be classified by their semantic domains. Examples of major classes of semantic domains are [Wing 90]:
- Abstract Data Type specification languages
- Process specification languages
- Programming languages

The distinction between operational and definitional methods provides another important dimension for classifying formal methods [Avizienis 90].

Operational methods have also been described as constructive or model-oriented [Wing 90]. In an operational method, a specification describes a system directly by providing a model of the system. The behavior of this model defines the desired behavior of the system. Typically, a model will use abstract mathematical structures, such as relations, functions, sets, and sequences. An early example of a model-based method is the specification approach associated with Harlan Mills' functional correctness approach. In this approach, a computer program is defined by a function from a space of inputs to a space of outputs. In effect, a model-oriented specification is a program written in a very high-level language. It may actually be executed by a suitable prototyping tool.

Definitional methods are also described as property-oriented [Wing 90] or declarative [Place 90]. A specification provides a minimum set of conditions that a system must satisfy. Any system that satisfies these conditions is functionally correct, but the specification does not provide a mechanical model showing how to determine the output of the system from the inputs. Two classes of definitional methods exist, algebraic and axiomatic. In algebraic methods, the properties defining a program are restricted to equations in certain algebras. Abstract Data Types are often specified by algebraic methods. Other types of axioms can be used in axiomatic methods. Often these axioms will be expressed in the predicate calculus. Edsger Dijkstra's method of specifying a system's (or process's) function by preconditions and post-conditions is an early example of an axiomatic method.

The use of formal methods will support knowledge modeling, learning, and reasoning of the underlying domain, for engineering system. To achieve that, it is essential to understand fundamentals of formal methods, which are based on predicate logic.

1.2 Logic

Logic or propositional calculus is based on statements, which have truth values (true or false). A proposition, or statement, is any declarative sentence, which is either true (T) or false (F). We refer to T or F as the truth value of the statement. The calculus provides a means of determining the truth values associated with statements formed from "atomic" statements. An example:

If p stands for "pressure is high in pipe P1" and q for "pipe P1 is leaking" then we may form statements such as shown in table 1-1.

Table 1-1. Symbolic Statements

Symbolic Statement	Translation
p ∨ q	p or q
p ∧ q	p and q
p ⇒ q	p logically implies q
p ⇔ q	p is logically equivalent to q
¬p (also ~p)	Not p

Note that ∨, ∧, ⇒, and ⇔ are all binary connectives. They are sometimes referred to, respectively, as the symbols for disjunction, conjunction, implication and equivalence. Also ¬ is unary and is the symbol for negation.

If propositional logic is to provide us with the means to assess the truth value of compound statements from the truth values of the 'building blocks' then we need some rules for how to do this. For example, the calculus states that "p ∨ q" is true if either p is true or q is true (or both are true). Similar rules apply for all the ways in which the building blocks can be combined. The language of predicate calculus requires: Variables and Constants. Table 1-2 shows first order predicate calculus.

Table 1-2. First Order predicate calculus

Symbol	Meaning
∨	or
∧	and
¬	not
⇒	logically implies
⇔	logically equivalent
∀	for all
∃	there exists

Quantification is non-logical constants that include names and entities. For example: ∀X.man(X) ⇒ mortal(X), means all men are mortal. ∃X.Tank(X), means there is at least one tank.

It is possible to form a new proposition from old one. For example, p: "There is Pump with 300 rpm in Plant Model Plant-1." The negation of p is ¬p, which is defined as: There is no Pump with 300 rpm in Plant Model Plant-1." Anther example: if p: "1 + 4 < 5", q: "1 + 4 = 5", then ~p∧~q: "1 + 4 > 5".

To generate new knowledge from old ones, one can use the truth table, which provides basic definition of predicate calculus. Table 1-3 (a), (b), and (c) shows the truth table for p, ¬p, p∨q, and p∧q.

Table 1-3. Truth Table of: (a) ¬p, (b) p∨q, and (c) p∧q.

(a) Negation

p	¬p
T	F
F	T

(b) Disjunction

p	q	p∨q
T	T	T
T	F	T
F	T	T
F	F	F

(c) Conjunction

p	q	p∧q
T	T	T
T	F	F
F	T	F
F	F	F

There are set of more complex expressions, that can be further simplified to validate the underlying system. Conditional and bidirectional are other forms of predicate calculus. For example, if p, then q can also be represented as: p implies q, and we write p⇒q. This can be represented as (¬p)∨q. Table 1-4 shows the truth table of such logic.

Table 1-4. (¬p)∨q.

p	q	p⇒q	¬p	(¬p) ∨q
T	T	T	F	T
T	F	F	F	F
F	T	T	T	T
F	F	T	T	T

Table 1-5. Examples of Conditional Statements

If p, then q. p implies q. q follows from p. Not p unless q. q if p. p only if q. Whenever p, q. q whenever p. p is sufficient for q. q is necessary for p. p is a sufficient condition for q. q is a necessary condition for p.

The biconditional p⇔q, which we read "p if and only if q" or "p is equivalent to q," is defined by the truth table, shown in table 1-6.

Table 1-6. Biconditional Statements.

p	q	p⇔q
T	T	T
T	F	F
F	T	F
F	F	T

These primitive statements and predicate calculus can be used for rules inference.

Modus Ponens or Direct Reasoning can be defined as: $[(p \Rightarrow q) \land p] \Rightarrow q$, which can be represented as:

p⇒q   (if p then q, or p implies q)
p   (p, which is true or false)

∴q

Similarly,
$(p \land q) \Rightarrow (r \land s)$
$(p \land q)$

∴(r∧s)

Modus Tollens or Indirect Reasoning can be defined as:
$[(p \Rightarrow q) \land \neg q] \Rightarrow \neg p$

In words, if p implies q, and q is false, then so is p.
p⇒q
¬q

∴¬p

Disjunctive Syllogism or One-or-the-Other:
[(p∨q) ∧ (¬p)] ⇒q
[(p∨q) ∧ (¬q)] ⇒p

Distributive Laws:
p∧ (q∨r)⇔(p∧q) ∨ (p∧r)
p∨ (q∧r)⇔(p∨q) ∧ (p∨r)

Applying Modus Ponens
1. (p∨q) ⇒ (r∧¬s)
2. ¬r⇒s
3. p∨q

Statement (p∨q) appears twice in lines (1) and (3). Looking at Modus Ponens, we see that we can deduce (r∧¬s) from these lines. Thus, we can enlarge our list as follows:

1. (p∨q) ⇒ (r∧¬s)   Premise
2. ¬r⇒s   Premise
3. p∨q   Premise
4. r∧(¬s)   1,3 Modus Ponens

Summary of Rules of Inference
- T1 Any tautology that appears on the list at the end of the last section can be used as a rule of inference.
- T2 We can add any tautology that appears in the list of tautologies at the end of the last section as a new line in our list of true statements.
- S (Substitution): We can replace any part of a compound statement with a tautologically equivalent statement.
- C (Conjunction): If A and B are any two lines in a proof, then we can add the line AB to the proof.
- P (Premise): We can write down a premise as a line in a proof.

1.2.1 Predicates

Upper case Roman letters, plus square brackets:
Examples:
H[a] : a is happy
R[a, b] : a respects b
S[a,b,g] : a sold b to g
H[a,b,g,d] : a is happy that b sold g to d

1.2.2 Function Signs

Lower case Roman letters, plus parentheses:
Examples:
m(a) : the mother of a
s(a,b) : the sum of a and b
s(a,b,g) : the sum of a, b, g
Variables: lower case Roman letters: z, y, x, …

1.2.3 Elementary Logic

Broadly speaking, logic is the study of good reasoning - and good reasoning is of considerable importance in many subjects. Elementary logic covers certain aspects of logic, which are regarded as fundamental. In elementary logic, compound terms, such as x + y, are not used where the focus is on the part of logic concerned with purely logical rules rather than with rules deriving from mathematical practice, or from some other specific domain. This restriction enables the use of a fairly simple syntax, and to make certain parts of the underlying logic practice fairly efficient.

It is recommended to stick to zero-order (i.e. propositional calculus) and first-order (predicate calculus) logic.

In general, elementary logic can include the following:

> sentences S
> noun phrases N
> predicates Nk$\rightarrow$S
> function signs Nk$\rightarrow$N
> connectives Sk$\rightarrow$S
> quantifiers V+S$\rightarrow$S
> description operator V+S$\rightarrow$N

1.3 Argument & Proofs

Precisely, an argument is a list of statements called premises followed by a statement called the conclusion.

An argument is a list of statements called premises followed by a statement called the conclusion.

P1 Premise
P2 Premise
P3 Premise
………..

Pr Premise

C Conclusion

The argument is said to be valid if the statement: $(P1 \wedge P2 \wedge \ldots \wedge Pr) \Rightarrow C$, is a tautology. In other words, validity means that if all the premises are true, then the conclusion must be true.

After having this quick highlights on logic or propositional and predicate calculus, the next sections shows aspects that are related to formal methods.

1.4 Automata Theory

Automata are abstract mathematical models of machines that perform computations on an input by moving through a series of states or configurations. Automata theory has close ties to formal language theory, since there is a correspondence between certain families of automata and classes of languages generated by grammar formalisms. A language is accepted by an automaton when it accepts all of the strings in the language and no others. The most restricted family of automata is finite automata consisting of only a finite number of states and a "read-only" tape containing the input to be read in one direction. Finite automata recognize the class of languages generated by regular (Type 3) grammars. These automata can be given a limited amount of extra power with the addition of certain forms of storage. For example, pushdown automata involve a pushdown store: a sequence in which symbols can only be added and removed from one end, with the effect that the first symbols in, are the last ones out. Pushdown automata accept the languages generated by context-free (Type 2) grammars. Automata theory gave rise to the notion of deterministic computation, hence deterministic languages. In a deterministic computation each configuration of the machine has only one possible successor. For some families of automata (e.g., finite automata and Turing machines) deterministic and nondeterministic automata are equivalent. For others (e.g., pushdown automata) there are languages that can be accepted by a non-deterministic automata of that family but cannot be accepted by any deterministic automata.

1.4.1   Deterministic finite state machine or deterministic finite automaton (DFA)

It is a finite state machine where for each pair of state and input symbol there is a deterministic next state.
A DFA is a 5-tuple, (S, Σ, T, s, A), consisting of:
A finite set of states (S)

A finite set called the alphabet (Σ)
A transition Quick Facts about: function
A mathematical relation such that each element of one set is associated with at least one element of another set function (T: $S \times \Sigma \to S$).
A start state (s . S)
A set of accept states (A . S)

Let M be a DFA such that $M = (S, \Sigma, T, s, A)$, and $X = x_1x_2 \dots x_n$ be a string over the alphabet Σ. M accepts the string X if a sequence of states, $r_0, r_1, \dots, r_n$, exists in S with the following conditions:
1. $r_0 = s$
2. $r_{i+1} = T(r_i, x_i)$, for $i = 0, \dots, n-1$
3. $r_n \square A$.

As shown in the first condition, that machine starts in the start state s. The second condition says that given each character of string X, the machine will transition from state to state as ruled by the transition function T. The last condition says that the machine accepts if the last input of X causes the machine to be in one of the accepting states. Otherwise, it is said to reject the string. The set of strings it accepts form a language, which is the language the DFA recognizes.

1.4.2 Nondeterministic finite state machine or nondeterministic finite automaton (NFA)

Is a finite state machine where for each pair of state and input symbol there may be several possible next states.
A NFA is a 5-tuple, (S, Σ, T, s, A), consisting of:
A finite set called the alphabet (Σ)
A finite set of states (S)
A transition Quick Facts about: function
A mathematical relation such that each element of one set is associated with at least one element of another set function ($T : S \times (\Sigma \ \square) \to P(S)$).
A start state (s $\square$ S)
A set of accept states (A $\square$ S) where P(S) is the power set of S and ε is the empty Quick Facts about: string A linear sequence of symbols (characters or words or phrases) string.

Let M be an NFA such that $M = (S, \Sigma, T, s, A)$, and X be a string over the alphabet Σ that can be written as $x_1x_2 \dots x_n$ where each xi $\square$ ($\Sigma \ \square$). M accepts the string X if a sequence of states,
$r_0, r_1, \dots, r_n$, exists in S with the following conditions:
1. $r_0 = s$

2. ri+1 □ T(ri, xi), for i = 0, ..., n-1
3. rn □ A.

The machine starts in the start state and reads in a string of symbols from its alphabet. It uses the transition relation T to determine the next state(s) using the current state and the symbol just read or the empty string. If, when it has finished reading, it is in an accepting state, it is said to accept the string, otherwise it is said to reject the string. The set of strings it accepts form a language, which is the language the NFA recognizes.

Every NFA has an equivalent DFA. Therefore it is possible to convert an existing NFA into a DFA for the purpose of implementing a simpler machine. This can be performed using the powerset construction, which may lead to an exponential raise in the number of necessary states.

1.4.3   Quick Facts about: Nondeterministic Finite Automata, with transitions (FND- or -NFA)

Besides of being able to jump to more (or none) states with any symbol, these can jump on no symbol at all. This is, if a state has transitions labeled with, then the NFA can be in any of the states reached by the -transitions, directly or through other states with -transitions. The set of states that can be reached by this method from a state q, is called the -closure of q. It can be shown, though, that all these automata can accept the same languages. You can always construct a DFA M that accepts the same language that a NFA M'.

1.5 Algorithms

In the context of formal methods, it is essential to brief the concepts of algorithms and its relation to problem solving. An algorithm is a sequence of simple steps that can be followed to solve a problem. These steps must be organized in a logical, and clear manner.

1.5.1 Algorithm Design

We design algorithms using three basic methods of control: sequence, selection, and repetition.

Sequential control

Sequential Control means that the steps of an algorithm are carried out in a sequential manner where each step is executed exactly once. Let's look at the following problem: We need to obtain the temperature expressed in Fahrenheit degrees and convert it to degrees Celsius. An algorithm to solve this problem would be: 1. Read temperature in Fahrenheit 2. Apply conversion formula 3. Display result in degrees Celsius.

In this example, the algorithm consists of three steps. Also, note that the description of the algorithm is done using a so-called pseudocode. A pseudocode is a mixture of English (or any other human language), symbols, and selected features commonly used in programming languages. Here is the above algorithm written in pseudocode:

- READ degrees_Farenheit
- Degrees_Celcius=(5/9)*(degrees_Farenheit-32)
- DISPLAY degrees_Celcius

Another option for describing an algorithm is to use a graphical representation in addition to, or in place of pseudocode. The most common graphical representation of an algorithm is the flowchart.

Selection Control

In Selection Control only one of a number of alternative steps is executed. Let's see how this works in specific examples.
Example: Control access to a computer depending on whether the user has supplied a username and password that are contained in the system database.

IF (user name in database AND password corresponds to user)
THEN Accept user
ELSE Deny user

Repetition

In Repetition one or more steps are performed repeatedly.
Example: Read numbers and add them up until their total value reaches (or exceeds) a set value represented by S.

WHILE (total<S)
DO Read number
        total=total+number
ENDDO

1.5.2 Problem Solving

Problem solving is to find set of actions to achieve the desired goals.

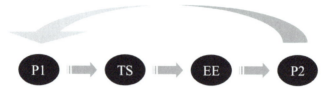

Figure 1-3. Problem Solving

Problem solving goes through several steps starting from defining the initial problem, generating tentative solutions, eliminate errors, then solve new problem till convergence or satisfaction.

P1 -- the initial problem
TS -- generate tentative solutions
EE -- eliminate errors
P2 -- the new problem

Problem solving process is to search for optimum solution in the problem domain. This can be achieved using the following steps:

- Define initial state
- Describe set of all possible actions, with state space and transition caused by these actions
- Define goals and performance indicators
- Define path cost function that assigns and calculates the cost for each possible path from initial state to final goal

Key feature in the problem solving process is the formal representation of the problem elements, i.e. initial state, actions, goals, and path costs. The search algorithm will greatly help to find the optimum solution with minimum cost.

1.6 Logic Programming

Knowledge can be represented in the form of semantic networks. Quine's web of belief (Quine, 1963, Quine and Ullian, 1970) shows the mind as a graph of connections (Kowalski 1975, 1979; Siekmann and Wrightson, 2002) among sentences expressed in a logical form. Sentences are in a special, simplified form, called clausal form; and the connections between sentences are also a simplified form of inference, called resolution [5]. Connection graphs have computational and logical interpretation. For example, activating a connection between a clause that represents a goal and a clause that represents a belief is a form of backward reasoning, which reduces the goal to sub-goals, similar to way in which a procedure call activates a procedure and invokes other procedure calls. Backward reasoning is the basic idea of logic programming (Kowalski, 1974) and the programming language Prolog.

Connection graphs can also simulate production systems. Activating a link between a clause that represents the record of an observation and a clause that represents a implicational goal is a form of forward reasoning (modus ponens), which derives a new goal, including the special case of a goal that is an atomic action.

In principle, Algorithm = Logic + Control.

In addition to logic programming, the term constraint functional logic programming was introduced.

The idea of Constraint Functional Logic Programming arose around 1990 as an attempt to combine two lines of research in declarative programming, namely Constraint Logic Programming and Functional Logic Programming. Constraint logic programming was started by a seminal paper published by J. Jaffar and J.L. Lassez in 1987, where the CLP scheme was first introduced. The aim of the scheme was to define a family of constraint logic programming languages CLP(D) parameterized by a constraint domain D, in such a way that the well established results on the declarative and operational semantics of logic programs could be lifted to all the CLP(D) languages in an elegant and uniform way. The best updated presentation of the classical CLP semantics can be found in [6]. In the course of time, CLP has become a very successful programming paradigm, supporting a clean combination of logic programming and domain-specific methods for constraint satisfaction, simplification and optimization, and leading to practical applications in various fields. On the other hand, functional logic programming refers to a line of research started in the 1980s and aiming at the integration of the best features of functional programming and logic programming. The first attempt to combine functional and logic languages was done by J.A. Robinson and E.E. Sibert when proposing the language LOGLISP [6].

Compositional logic programming is a combinator is a predicate that takes one or more partially applied predicates as arguments. A compositional logic programming language allows programs to be constructed using both the primitive connectives of the language and user-defined combinators such that the termination behavior of the program is independent of the way in which it is composed.

1.6.1 Example

A Prolog program is a set of clauses (logical sentences) written in a subset of first-order logic called Horn clause logic, which means that they can be interpreted as if–statements. A predicate is a set of clauses that defines a relation, i.e. all the clauses have the same name and arity (number of arguments).
Predicates are often referred to by the pair name/arity. For example, the predicate in_tree/2 defines membership in a binary tree:

in_tree(X, tree(X,_,_)).
in_tree(X, tree(V,Left,Right)) :- X<V, in_tree(X, Left).
in_tree(X, tree(V,Left,Right)) :- X>V, in_tree(X, Right).

(Here ":- " means if, the comma "," means and, variables begin with a capital letter, tree(V,L,R) is a compound object with three fields, and the underscore "_" is an anonymous variable whose value is ignored.) In English, the definition of in_tree/2 can be interpreted as: "X is in a tree if it is equal to the node value (first clause), or if it is less than the node value and it is in the left subtree (second clause), or if it is greater than the node value and it is in the right subtree (third clause)."

1.6.2 Dynamic typing

Compound data types are first class objects, i.e. new types can be created at run-time and variables can hold values of any type. Common types are atoms (unique constants, e.g. foo, abcd), integers, lists (denoted with square brackets, e.g. [Head|Tail], [a,b,c,d]), and structures (e.g. tree(X,L,R), quad(X,C,B,F)). Structures are similar to C structs or Pascal records—they have a name (called the functor) and a fixed number of arguments (called the arity). Atoms, integers, and lists are used also in Lisp.

1.6.3 Unification

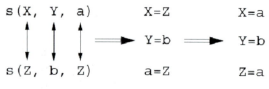

Figure 1-4. Unification Example

Unification is a pattern-matching operation that finds the most general common instance of two data objects. A formal definition of unification is given by Lloyd [42]. Unification is able to match compound data objects of any size in a single primitive operation. Binding of variables is done by unification. As a part of matching, the variables in the terms are instantiated to make them equal. For example, unifying s(X,Y,a) and s(Z,b,Z) (Figure 1-4) matches X with Z, Y with b, and a with Z. The unified term is s(a,b,a), Y is equal to b, and both X and Z are equal to a.

1.7 Formal Languages

In general languages can be split up into two main groups: natural languages, such as French, English and German, and formal languages (or computer languages), such as Pascal, COBOL and FORTRAN. Formal languages are formed from symbols, strings, and alphabets. Symbol can be one of the following:

- One of the digits 0-9;
- One of the upper case letters A-Z;
- One of the lower case letters a-z;
- A (, followed by any characters, provided that "(" and ")" are properly nested.

Formal language F is specified by:
(1) The vocabulary of F
(2) The rules of formation of F

1.7.1 Formal Language Anatomy

Vocabulary
(1) Upper case Roman letters: A B C ......
(2) Nothing else is a symbol of F.

Formation Rules
(1) If a string STR begins with 'P', then it is a formula.
(2) Nothing else is a formula of F.

1.7.2 Object Language versus Meta-Language

Consider French; there are two types of French grammar books; those written in French, and those written in a "foreign" language (say, English). In both cases, French is the subject of discussion (or object of discussion). However, whereas the former book is also written in French, the latter book is written in

English. In both cases, French is the language under discussion, the object language. What about the language of use? The meta-language? Well, in the former case, French is also the language of use (the meta-language); in the latter case, English is the language of use.

1.7.3   Formal Language Examples

1.7.3.1   Abstract State Machines

Abstract State Machine is a formal method for specification and verification. The Abstract State Machine (ASM) Project (formerly known as the Evolving Algebras Project) was started by Yuri Gurevich as an attempt to bridge the gap between formal models of computation and practical specification methods.

To write a program in a language like C or Java, various statements are used, such as: conditional statements, loop statements, and so forth. ASM statements are called rules. The most basic rule7 is the update rule, which has the form:
foo(t1, t2,. . . tn) := t0

Here, foo is an n-argument function, and t0 through tn are terms, or expressions. Executing the rule updates the value of the function foo at the specified arguments to the specified value.

1.7.3.2   B-Method

The B-Method is a collection of mathematically based techniques for the specification, design and implementation of software components. Systems are modeled as a collection of interdependent Abstract Machines, for which an object-based approach is employed at all stages of development.

An Abstract Machine is described using the Abstract Machine Notation (AMN). A uniform notation is used at all levels of description, from specification, through design, to implementation. AMN is a state-based formal specification language in the same school as VDM and Z. An Abstract Machine comprises a state together with operations on that state. In a specification and a design of an Abstract Machine the state is modelled using notions like sets, relations, functions, sequences etc. The operations are modelled using Pre- and Post-conditions using AMN.

In an implementation of an abstract machine the state is again modeled using a set-theoretical model, but this time we already have an implementation for the model. The operations are described using a pseudo-programming notation that is a subset of AMN.

The B-Method prescribes how to check the specification for consistency (preservation of invariant) and how to check designs and implementations for correctness (correctness of data refinement and correctness of algorithmic refinement).

The B-Method further prescribes how to structure large designs and large developments, and promotes the re-use of specification models and software modules, with object orientation central to specification construction and implementation design.

A great deal of attention has been paid to making the notational aspect of the method as simple as possible. To the engineer, the formal notation looks like a simple pseudo programming notation. And as mentioned above, there is no real distinction between the specification notation and the programming notation.

1.7.3.3   Petri Nets

The concept of Petri nets has its origin in Carl Adam Petri's dissertation Kommunikation mit Automaten, submitted in 1962 to the faculty of Mathematics and Physics at the Technische Universität Darmstadt, Germany.

A Petri net is a graphical and mathematical modeling tool. It consists of places, transitions, and arcs that connect them. Input arcs connect places with transitions, while output arcs start at a transition and end at a place. There are other types of arcs, e.g. inhibitor arcs. Places can contain tokens; the current state of the modeled system (the marking) is given by the number (and type if the tokens are distinguishable) of tokens in each place. Transitions are active components. They model activities, which can occur (the transition fires), thus changing the state of the system (the marking of the Petri net). Transitions are only allowed to fire if they are enabled, which means that all the preconditions for the activity must be fulfilled (there are enough tokens available in the input places). When the transition fires, it removes tokens from its input places and adds some at all of its output places. The number of tokens removed / added depends on the cardinality of each arc. The interactive firing of transitions in subsequent markings is called token game.

Petri nets are a promising tool for describing and studying systems that are characterized as being concurrent, asynchronous, distributed, parallel, nondeterministic, and/or stochastic. As a graphical tool, Petri nets can be used as a visual-communication aid similar to flow charts, block diagrams, and networks. In addition, tokens are used in these nets to simulate the dynamic and concurrent activities of systems. As a mathematical tool, it is possible to set up state equations, algebraic equations, and other mathematical models governing the behavior of systems.

1.8 Conclusion

This chapter presented essential basics about formal methods where reader can understand fundamental concepts of logic, logic programming, formal languages, automata theory, problem solving methods, and recognize some examples of known formal languages.

1.9 References

[1] Tom M. Mitchell. Machine Learning. ISBN 0-07-042807-7.
[2] Hossein Saiedian, Michael G. Hinchey (1996). Challenges in the successful transfer of formal methods technology into industrial applications. Information and Software Technology, Vol. 38 (1996), 313-322.
[3] Robert L. Vienneau (1993). Data & Analysis Center for Software, New York, http://www.dacs.dtic.mil/techs/fmreview/title.html.
[4] Egidio Astesiano, Gianna Reggio (2000). Formalism and method. Theoretical Computer Science, Vol. 236, 3-34.
[5] Robert Kowalski. Logic and Modules. Technical Notes, Department of Computing, Imperial College London, Apr-2005. http://www-lp.doc.ic.ac.uk/UserPages/staff/rak/papers/Modularity.pdf.
[6] F. Javier López-Fraguas, Mario Rodríguez-Artalejo and Rafael del Vado Vírseda (2005). Constraint Functional Logic Programming Revisited. Electronic Notes in Theoretical Computer Science, Volume 117, 20 January 2005, Pages 5-50.
[7] Hardegree (2003). Metalogic and Formal Languages. http://people.umass.edu/gmhwww/595/pdf/metatheory/MetaTheory-01-Formal%20Languages%201.pdf.
[8] James K. Huggins, Charles Wallace (2002). An Abstract State Machine Primer. http://www.eecs.umich.edu/gasm/
[9] Georges Mariano (2004). The B Formal Method Bibliography. http://download.gna.org/brillant/docs/B-Bibliography/B-Bibliography.pdf.

2 Formal Methods for Process Systems Engineering

Author

Hossam A.Gabbar,
Graduate School of Natural Science & Technology,
Okayama University

Summary

This chapter presents formal methods and approaches for process systems engineering. This requires giving a brief introduction on process systems engineering and the difficulties and challenges and how formal methods can address these difficulties.

Among the challenges that faces process systems engineering is the process design and plant operation engineering and management. This chapter will provide practical formal methods approaches that are useful for process design and plant operation practices.

Keywords: SOP synthesis; Operating Procedures Synthesis; Operation Engineering; EFL; Engineering Formal Language

2.1 Introduction

Many researchers around the world in the area of chemical engineering have recognized the lack of practical automated solutions that can adequately support life cycle activities of chemical processes. The problem has been initiated from two angels, (1) when trying to develop automated solutions to support process design and operation, there was a gap between process models and views and systems models, which led to the unrealistic computer-aided engineering solutions, (2) when trying to analyze life cycle activities to improve process design and operation, there was lack in the available process modeling techniques, and many researchers tried to use system modeling concepts, which were not satisfactory for process engineering views..

Process Systems Engineering is becoming increasingly popular with engineers and scientists serving the process industries. It includes useful practices for academia and industry as well as challenges and future trends, and innovative work in all aspects of process systems engineering, which includes (but not limited to): modeling and simulation methodologies, process / product design, process control, plant operation, plant maintenance, planning, optimization,

Hossam A. Gabbar (ed.), Modern Formal Methods and Applications, 21–35.
© 2006 *Springer. Printed in the Netherlands.*

manufacturing execution systems, production management, supply chain, hybrid dynamic systems plant maintenance, process safety, environmental assessment and risk management. In process systems engineering domain nano, micro, with macro levels are integrated smoothly to support life cycle activities using mathematical and artificial intelligence practices. In such complex domain, the use of knowledge engineering, expert system, and advanced technologies of multimedia and augmented reality will be investigated and applied to provide better computer-aided process engineering solutions.

2.2 Process Systems Engineering

Process Systems Engineering is becoming increasingly popular with engineers and scientists serving the process industries. There are many challenges that face plant and product life cycle, which requires innovative process engineering practices including (but not limited to): modeling and simulation methodologies, process / product design, process control, plant operation, plant maintenance, planning, optimization, manufacturing execution systems, production management, supply chain, hybrid dynamic systems plant maintenance, process safety, environmental assessment and risk management. There are new innovative solutions and ideas that integrate nano, micro, and macro level processes to better support life cycle activities. In such complex domain, the use of knowledge engineering, expert system, and advanced technologies will be the cornerstone to provide better computer-aided process engineering solutions.

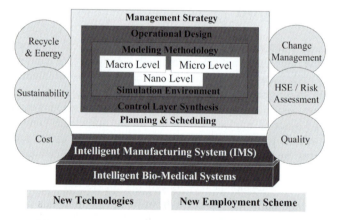

Figure 2-1. Process Systems Engineering Roadmap

Figure 2-1 shows layers of process systems engineering, which includes modeling and simulation of nano, micro, and macro levels. The operation and control practices are necessary for safe, optimum, and steady production, which will be achieved through management, planning and scheduling. Cost, quality, health, risk, and other factors will be evaluated throughout life cycle activities.

2.3 Why Formal Language?

Formal representation provides a systematic framework to construct and validate the syntax of the underling system towards building standard representation approaches. Formal languages are used in most of the engineering applications to reduce the time and efforts required to communicate and manage the underling system or process. Formal language implies absolutely accurate and precise definitions of the underling system, which can be used to validate the system / process and can be used as a base for computer-aided solutions. One can say that the use of formal languages will help overcoming various engineering problems.

Engineering problems are always referred to performance, optimization, risk, user friendliness, cost effective, etc. These factors are always studied using conceptual or mathematical modeling. However, both ways of modeling approach requires detailed domain knowledge which is always incomplete, or there is no formal way to ensure its continuation and tuning throughout the progress of process / product life cycle.

The construction of domain knowledge can be systemized if a robust model formalization methodology is used. Many researchers have investigated the construction of process and plant models, and proposed modeling methodologies to build the corresponding domain knowledge. Most of these modeling methodologies showed that plant model includes three views: structure, behavior, and operation (Gabbar et al., 2000; Lu et al., 1995, 1997).

Among the different challenges that face engineering systems is the automatic and accurate synthesis of operating procedures, especially when having large chemical plants with complex plant operation. Operating procedures are sets of tasks / activities that are carried out to produce the target products using input materials and the available topological resources. Researchers viewed the automation of operating procedures synthesis from different angles. One key view is during process design where operation is considered while defining plant structure and selecting behaviors and topology paths. Another view is product design and conversion process from input raw materials to producing the final output. One more view is the operation optimization which is achieved through thorough design of plant operation.

In all these views, there is a need to systematically define operating procedures and represent them in a way that both machine and human can understand. This leads us to the use of formal methods with simplified representation. Formal representations of operating procedures facilitate engineering of plant operation and also human computer interaction. From the other side, formal representation will provide means to build and organize domain knowledge more efficiently.

Many researchers are motivated to use computer languages such as HTML, XML, or even build their own languages on the basis of these computer languages to describe their systems or processes. For example, world batch forum proposed the use of XML to represent batch recipe (WBF). However, it is essential to define engineering language in a higher layer, which can deal with the syntax and semantics of operating procedures as linked to plant model (structure and behavior), before addressing the implementation issues using computer languages such as XML, HTML, etc. It will be practical if two separate layers are defined to represent operating procedures. The higher layer will enable designers and engineers to define operating procedures in standard and formal way as mapped to the underling system, regardless of the implementation way. This will enable them to think freely without the limitations of any computer language. Based on the higher layer, the lower layer can be designed or selected for implementation. In general, the design of operating procedures requires flexible way to define and maintain the syntax and semantics of operating procedures as it involves different participants with different views. It is relatively easy to maintain the syntax of the defined higher layer language, as it doesn't involve modifications to computer languages and programs. There could be mapping or translation from high level formal language into computer language, which is relatively easy and can be implemented for any destination (i.e. host) computer language in the lower layer (e.g. HTML, XML, C, etc.).

As described by Carre' (1989) and reported by Toyn (1998), formal language should be logically sound and unambiguous and the semantics must not be too complex, otherwise formal reasoning will become impractical.

The proposed engineering formal language is based on the definition of simple and clear English-like statements, which are composed of keywords that are mapped to domain knowledge. The structure of the statements defines the syntax while the mapping to domain knowledge defines the semantics. The proposed engineering formal language will fulfill the basic requirements of any formal language where hierarchical definitions are used to simplify the representation of complex expressions. For example, one can say: "valve is active-structure", "pump is active-structure", while "tank is passive-structure". Also, "tank" can be further explored as "tank has input-ports" and "tank has output-ports".

As a conclusion, the use of formal representation will facilitate the design and validation of engineering systems in structured way as mapped to domain knowledge. In addition, the use of formal language will facilitate the development of automated computer-aided engineering solutions for the advancement of process engineering.

2.4 Operation Engineering

Within process engineering practices, formal methods can be used in different applications such as process design, maintenance, and operation engineering. This section will describe how formal methods could be used in operation engineering.

Plant operation goes through different stages and activities from conceptual design, detailed design, till engineered design. In fact it is similar to process design, which goes also through different stages such as conceptual design, detailed design, etc.

Operation engineering requires complete understanding of process design and the how to produce the target product(s). In addition, it is essential to understand the operational aspects of each structure unit and control requirements. There are different challenges that face operation engineering such as operating procedures structuring, operating procedure representation, process design knowledge structuring, and operation optimization. In addition, it is essential to understand the different hazardous situations, safety constraints, and regulations, and the counteractions that need to be considered to ensure process safety throughout plant operation.

Operating procedures synthesis is quite complex process which can be viewed in different levels of complexities. SOP or standard operating procedures usually include all engineering aspects of the underlying plant, which satisfy three major targets: safe, optimum, and steady operation.

The structuring and representation of SOP is quite important to facilitate the operation understanding, automatic generation, and validation of operating procedures. In the next section, structure of SOP will be presented.

2.5 SOP Synthesis
2.5.1 SOP Structure

Operation is viewed as sequence of actions / tasks that require resources such as materials, equipments, human, etc. There are useful standards that tried to provide unified structure for operating procedures. For batch recipe, ANSI/ISA-S88 proposed hierarchical structure that links plant structure with plant operation. Figure 2-2 shows the proposed structure by ANSI/ISA-S88, where operating procedures are classified as procedure, unit procedure, operation, and phase. These operation levels are mapped to plant structure hierarchy, which shows: cell, unit, and equipment module. Operating procedures are classified as general recipe, site recipe, master recipe, and control recipe. Each level can be defined in terms of procedure, unit procedure, operation and phase.

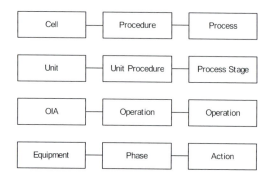

Figure 2-2. ANSI/ISA-S88 Operation Model

The automatic synthesis of SOP requires systematic definition and representation of plant structure hierarchy and operation domain and structured formal representation using domain knowledge.

Each procedure, unit procedure, operation, and phase can be represented as a set of tasks. Each task includes action, pre-condition and post-condition. To systematically construct SOP, it is essential to find a way to synthesize operation actions, pre-condition, and post-condition. Each task can be structured following a standard syntax such as:
- ✓ Action
- ✓ Pre-condition
- ✓ Post-condition

Each of these three components of SOP task can be constructed, represented, and validated using formal methods and simplified engineering formal language. The proposed engineering formal language (or EFL) consists of vocabulary and formal rules. Vocabulary is formed from domain knowledge, while formal rules are constructed from process constraints and control rules.

2.5.2 Tokens

In this section, a systematic method will be illustrated to extract the vocabulary of EFL. Vocabulary is composed of keywords or tokens, which are extracted usually from domain knowledge. There is a semantic meaning of these tokens and there are some relationships among them, which can be described using ontological engineering. The extracted tokens are classified in a way to support the construction of SOP using EFL. In addition, such token classification will enable the validation of the generated (or synthesized) SOP.

Table shows examples of the identified tokens from domain knowledge.

Table 2-1. Examples of Tokens within EFL

MOVE	Constant
Equip-id	Lookup
Material-id	Lookup
Role-id	Lookup
upper-limit	Input
Transportation-Actions	Variable
Transformation-Actions	Variable

In such table, "Constant" is used to define tokens that are used as they are within EFL. "Lookup" is used to show tokens that are used to link EFL with list of options or alternatives within corresponding database. "Input" is used when user is required to provide some inputs. And "Variable" is used when the token will be further explained using another EFL statement.

2.5.3 Domain Knowledge

As part of the requirement analysis and process system engineering practices, domain knowledge is modeled; this includes all concepts and their classifications.

Ontology modeling can be used to construct such operation hierarchies and can be used to define the set of tokens, constraints, and conditions associated with each operation. The association between operation, behavior, and structure can also be represented within the proposed ontology model. Figure 2-3 shows parts of the developed ontology model within ontology editor. The developed ontology model is used as a base to construct EFL statements and to validate the semantics of operating procedures tasks. Currently, base ontology model has been developed using ontology editor and converted into database repository, which is used by operating procedures synthesis automated solutions to design plant operation.

In case of batch plants, master and control recipe statements are defined on the basis of EFL where keywords are linked to domain knowledge such as plant design model, material, functions, products, recipe formula, etc. For example, master recipe statement (MOVE material1 FROM t1 TO t2) is derived from EFL statements S1 & S2.

Transportation_Action :: MOVE material FROM Topology_Area TO Topology_Area
[S1]

Topology_Area :: cell_id | unit_id | oia_id | ([class_function]
[class_material] equip_class) | equipment_id
[S2]

Where material1 is selected from a lookup list of all materials defined within the domain knowledge of the used plant model. "FROM" and "TO" are constant keywords, while t1 and t2 are equipment id's, which are selected from lookup list of all equipments defined in the plant model.

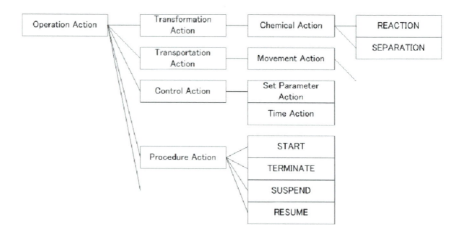

Figure 2-3. Operation Ontology

2.5.4   Formulas

Within operating procedures, formulas are mathematical equations that are used to calculate the quantitative amounts associated with each SOP task. There are different ways to formalize formulas. EFL can be used to systematically structure formulas where basic EFL statements are constructed and used to construct more complex formulas. For example, the formula "$x = y^2 + z*a$" can be constructed using "variable is (expression operator expression) | variable". There are well proven simulation solvers, which are based on robust mathematical representations that can be used to solve very complex equations.

2.5.5 Synthesis of Meta-Operation

In case of batch plants, and in the course of defining master recipe of batch plants, there are different tasks that are repeated in different operations. For example, cleaning operation can be applied to different structure units in different unit procedures such as reaction, separation, etc. The description and detailed tasks of the cleaning operation are the same, while the associated values, condition, and plant structure might be different. Providing complete and generic description of such repeated operations will reduce the volume and complexity of operating procedures and will enable operation designers to describe and build complex operations from simplified smaller operation tasks (i.e. libraries or modules). Repeated and generic operations will be called meta-operation, which can be identified by experts who are knowledgeable of plant design and operations. This process (i.e. building meta-operation libraries) is time consuming and can be considered as an ongoing process, which might involve teams from different departments / sections / enterprises to iteratively tune and modify meta-operations. In current situation, teams and individuals use different operation formats, terminologies, and tools, which made it difficult to work together in one format and talk the same language. This can be improved by providing automated environment to manage such iterative process and to enable the different teams to maintain their operating procedures as well as meta-operation in standard format. EFL can provide a standard format and unified language, which can be used as a base to provide automated solution to synthesize operating procedures. The proposed automated solution includes EFL editor and parser, which will enable the synthesis of standard operating procedures and will facilitate the exchange of operation libraries and meta-operations in neutral format i.e. EFL.

2.5.6 Examples of Meta-Operation
2.5.7 Isolation

Isolation is one example of common operations, which might be used when maintenance work is required, or as a recovery operation when process equipment encounters a failure. Kim (2000) and Asprey (1999) showed some examples to synthesize safe operation. Kim (2000) used SMV or symbolic model verifier to find errors and to synthesize safe operation, while Asprey (1999) used simulation based approaches to avoid unsafe situations. Both proposed useful ideas to synthesize safe operation, which requires robust mechanism to represent recovery operation.

Isolation, as a recovery operation, has the generic meaning of closing all in and out connections around the underling process equipment or topology area. This is equivalent to closing all control valves surrounding that process equipment. The isolation might be partial where only upstream or downstream control devices are required to be closed.

2.5.8 Cleaning

Cleaning is another example, which might be included in different operations and executed in different topology areas. There are different scenarios that can decide the way to perform cleaning operation. For example, fluid type (i.e. gas or liquid), pressure and temperature levels can decide how to perform cleaning operation. From the underling domain knowledge, fluid type and method of cleaning can be decided. For example, high-pressure gas, high flow rate water, or hot water could be used to clean specific tank. During the design stage, cleaning method might be defined for some structure units as part of process design model. MDOOM (Lu et al., 1995; 1997) and POOM (Gabbar et al., 2000; 2004; 2005) are process modeling methodologies, which show three views to describe process design: structure/static, behavior/dynamic, and operation/function. Operation related information about the cleaning operation might be associated with structure units as part of the operation view. For example tank "T1" might be associated with cleaning method, which has water as cleaning fluid. This will dictate the selection of the suitable cleaning operation.

Cleaning operation can be described in generic form as: move cleaning-fluid from cleaning-fluid-source to cleaning-fluid-destination. Such generic cleaning operation can be assigned to a given topology area, which will define the type of cleaning fluid and will specify the source and destination topology areas. Semantic errors can be discovered using the defined domain knowledge. For example, if there is no water source for tank "T1", then such water-cleaning operation cannot be performed.

2.5.9 Heating

Another example of the use of meta-operations is the heating operation, which is commonly used in chemical processes. Heating operation can be performed using heating source in the underling structure unit. Heat source could be hot air, fuel, hot water, etc. Similar to the cleaning operation, operation-related information might be stored as associated with the structure unit, i.e. type and name of heating fluid. The generic form of heating operation could be in the form: supply heating source to the specified structure unit. This operation might have some parameters such as time for heating, temperature, etc. Some of this information will be associated with the control devices (servo or regulatory controllers) associated with the underling structure unit.

Cooling operation can be represented similarly where cooling source will be used instead of heating source.

2.5.10 Recovery

There are different ways to realize recovery operation, based on the topology area and type of failure as well as fault propagation scheme. Some of the recovery operations can be realized using isolation meta-operation. For example, In case of leak (or other abnormal situations), close all upstream valves associated with such unit (e.g. tank). This requires generic precondition, which can be described as "if abnormal situation occurred".

The following section describes the proposed mechanism to construct and utilize meta-operations for a case study batch plant using RFDL.

2.6 Meta-Operation for Master Recipe

Meta-operation is a useful technique to optimize master recipe where repeated operations can be generalized in the form of meta-operation. This will enable operation designers to define sorts of operation libraries, which will be used with different plant structures and unit procedures. This will reduce the time and effort to construct master recipe and will reduce the errors where meta-operations will be defined once and used in different places. Meta-operation will have id and description. While creating master recipe, once user selects one meta-operation from the meta-operation list (i.e. using the id or description), the associated definition (i.e. in the form of EFL) will be copied to that master recipe operation task. As per the definition of meta-operation, the initiation & termination triggers as well as action will be defined for all elements of the corresponding operation level task (i.e. initiation trigger of meta-operation is copied to the initiation of the operation level task, and same for termination trigger and action). As mentioned in the above cooling example, one meta-operation could be "move cooling-fluid from cooling-fluid-source to cooling-fluid-destination in topology-area". This might have a generic precondition, which should be defined during the selection of that meta-operation in master recipe. The generic form of precondition of meta-operation is in the form "i_precondition", or "i_postcondition", or "i_action". The master recipe editor will understand that such generic task element should be defined during the selection of the meta-operation in master recipe.

Meta-operation is used mainly with the operation level master recipe (i.e. for specific unit procedure), however, it can also be used with unit procedure. In such case, meta-operation will be defined as two levels meta-operation (i.e. meta-unit-procedure and meta-operation underneath). For example, user can define initial setting as one meta-unit-procedure, where it includes isolation, initial cooling, and isolation meta-operations. Once user selects such meta-unit-procedure, system will create one unit procedure as "initial setting", which includes three operations: "isolation", "initial cooling", and "isolation". Each will have set of generic tasks associated with precondition, postcondition, and

actions. Table 3 shows detailed structure of meta unit-procedure, which
includes three meta-operations. One user selects such meta-unit-procedure, all
meta-operation will be selected as well. User will be prompted to define values
for the identified parameters.

Table 2-2. Meta-operation Example for Master Recipe

Meta-Operation: id: MO-UP001, Description: Initial Setting, Type: [Unit
Procedure], Parameters: (Unit)
Task-1
Precondition: i_precondition
Postcondition: i_postcondition
Action: i_action

Meta-Operation: id: MO-UP001-OP001, Description: Isolation, Type:
[Operation], Paramters: (Topology_Area)
Task-1
Precondition: i_precondition
Postcondition: i_postcondition
Action: CLOSE ALL VALVES OF Topology_Area

Meta-Operation: id: MO-UP001-OP002, Description: Initial Cooling, Type:
[Operation], Paramters: (Topology_Area)
Task-1
Precondition: i_precondition
Postcondition: i_postcondition
Action: MOVE cooling-fluid FROM cooling-fluid-source TO cooling-fluid-
destination OF Topology_Area

Meta-Operation: id: MO-UP001-OP003, Description: Isolation, Type:
[Operation], Paramters: (Topology_Area)
Task-1
Precondition: i_precondition
Postcondition: i_postcondition
Action: CLOSE ALL VALVES OF Topology_Area

2.7 Control Recipe Generation

The corresponding control recipe of the defined master recipe of "CLOSE ALL
UPSTREAM VALVES OF AJ4101" can be viewed in figure 2-4. It shows the
close of all upstream valves of the tank AJ4101 (both the jacket and the inner
tank).

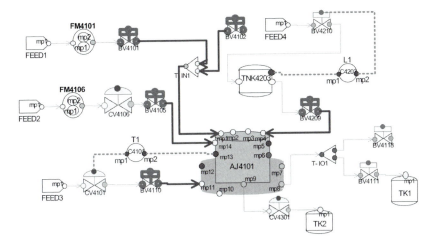

Figure 2-4. Visualization of Isolation Meta-operation

2.8 Conclusion

Formal methods can be used effectively to synthesize and validate operating procedures. Engineering formal language, or EFL, is proposed to construct master recipe statements and synthesize the corresponding control recipe. For systematic construction of operating procedures, meta-operation is introduced where it is used to define group of operations that are likely to frequently occur. Meta-operation is used to describe generic operations of chemical batch plants, which can also be used in continuous and discrete manufacturing / production plants. EFL or engineering formal language is used to represent meta-operation (and operating procedures in neutral and standard format, which enable the synthesis of operating procedures. Batch plant is used as a case study where master recipe is synthesized using EFL. Case studies are used from chemical batch plant to show the proposed synthesis and representation mechanisms of meta-operations.

The proposed solution will enable operation designers to synthesize meta-operations, which can be used while defining master recipe as well as control recipe of chemical batch plants. The proposed EFL and representation mechanism will enable the smooth and accurate generation of SOP of any complex chemical plants. In addition, it can be used efficiently to validate and visualize SOP.

Further research work is required to define a systematic mechanism to define hierarchical and more complex meta-operations for complex plants. Also it is required to define generic conditions (i.e. precondition and postcondition), which could be linked to product quality, batch planning, and plant maintenance details.

2.9 References

ANSI/ISA-S88.01, 1995. Batch Control. Part 1. Models and terminology.

Aoyama, A., Yamada, I., Batres, R., & Naka, Y. (2000). Development of batch process operation management platform. The 10th European Symposium on Computer Aided Process Engineering. Special Issue of Computers & Chemical Engineering, 24, 519-524.

Arzen, K.E. (1994). Grafcet for intelligent supervisory control applications. Automatica 30(10), 1513-1525.

Asprey, S., Batres, R., Fuchino, T., and Naka, Y. (1999). Simulation-based operations planning in the presence of quantitative safety constraints. Proceedings of 2nd Conference on Process Integration, Modeling and Optimization for Energy Saving and Pollution Reduction (PRES'99), Budapest, Hungary, pp.133-138 (1999).

Gabbar, H.A., Aoyama, A., Naka, Y. (2003). Model-Based Computer-Aided Design Environment for Operational Design. Journal of Computers & Industrial Engineering (Submitted).

Gabbar, H.A., Aoyama, A., Naka, Y. (2003). AOPS an Automated Solution for Operating Procedures Synthesis for Batch Plants. Journal of Computers & Chemical Engineering (Submitted).

Gabbar, H.A., Aoyama, A., Naka, Y. (2003). Recipe Formal Definition Language For Operating Procedures Synthesis. Journal of Computers & Chemical Engineering (Submitted).

Gabbar, H.A., Chung, P.W.H., Suzuki, K., and Shimada, Y. (2000). Utilization of unified modeling language (UML) to represent the artifacts of the plant design model. Proceedings of "PSE Asia 2000" International Symposium on Design, Operation and Control of Next Generation Chemical Plants, PS54, 387-392, Kyoto-Japan.

Gabbar, H.A. and Naka, Y. (2003). Computer-Aided Operation Design Environment for Chemical Production Plants. ICCTA'2003 – IEEE, 12th International Conference on Computer Theory and Applications, Aug-2003, Alexandria, Egypt, P27.

Johnsson, C. and Arzen, K.E. (1998). Grafchart for recipe based batch control. Computers & Chemical Engineering, 22, 1811-1228

Kim, J. and Moon, I. (2000). Synthesis of safe operating procedure for multi-purpose batch processes using SMV. Computers & Chemical Engineering, Vol. 24 (2000), Issues 2-7, 385-392.

Kim, M.and Lee, I.B. (1997). Rule-based reactive rescheduling system for multi-purpose batch processes. Computers & Chemical Engineering, Vol. 21, Suppl. Pp. S1197-S1202.

Kirkwood, R.L., Locke, M.H., and Douglas, J.M. (1988). A prototype expert system for synthesizing chemical process flowsheets. Computers & Chemical Engineering, Vol. 12 (1988), Issue 4, 329-343.

Lakshmanan, R. and Stephanopoulos, G. (1990). Synthesis of operating procedures for complete chemical plants – I Hierarchical, structured modelling for nonlinear. Computers & Chemical Engineering, Volume 14 (1990), Issue 3, 301-317.

Lu, M. L., Batres, R., Li, H. S., and Naka, Y. (1997). A G2 based MDOOM testbed for concurrent process engineering. Computers & Chemical Engineering, Vol. 21, Suppl., pp. S11-S16.

Lu, M. L., Naka, Y., Shibao, K., Wang, X. Z., and McGreavy, C. (1995). A multi-dimensional object-oriented model for chemical engineering. Proceedings of the Second International Conference on Concurrent Engineering, Research and Application, Virginia, Aug. 1995, USA, pp. 21-29.

Naka, Y. and McGreavy, C. (1994). Modular approach for startup operational procedures of chemical plant. Proceedings of PSE'94, 1007-1013.

Naka, Y., Batres, R., and Fuchino, T.; Operational design and its benefits in real-time use, Foundations of computer aided process operations (1999), ISBN 0-8169-0776-5, pp: 570.

Ruiz, D., Canton, J., Nougues, J.M., Espuna, A., and Puigjaner, L. (2001). On-line fault diagnosis system support for reactive scheduling in multipurpose batch chemical plants. Computers & Chemical Engineering, Vol. 25, pp. 829-837.

Viswanathan, S., Johnsson, C., Srinivasan, R., and Venkatasubramanian, V., & Arzen, K.E. (1998). Automating operation procedure synthesis for batch processes: Part I. Knowledge representation and planning framework. Computers & Chemical Engineering, 22, 1673-1685.

3 Formal Methods for Production Chain Management

Author
Hossam A.Gabbar,
Graduate School of Natural Science & Technology,
Okayama University

Summary
To meet the dynamically changing market requirements, production enterprises are collaborating in the form of production chains to improve production efficiency and product quality in view of the dynamic market changes and demands. In order to overcome the difficulties of operating such heterogeneous enterprises and to link the operation of different hierarchical levels within each enterprise, a robust operation representation framework is required. This research paper presents a formal representation approach for production chain operation, which ensures unified operation between the micro level, i.e. process, and macro level, i.e. production line, of enterprises within production chain. The proposed formal methods are used to represent operating procedures, process constraints, and control rules. Case study production chain is used to illustrate the proposed operation representation approach.

Keywords: formal representation, production chain operation, operation representation, SOPFL, ODM

3.1 Introduction

To meet the dynamically changing market requirements it is essential to provide flexible manufacturing enterprises, which will be able to dynamically tune their production systems as per market requirements and social demands. To achieve such level of dynamic or agile manufacturing, it is required to reduce the gap between enterprise management, production, and process levels, as well as between suppliers and customers in the so-called production chains. Production chains are networks of enterprises, service providers, clients / customers where raw materials and services are transformed from provider / manufacturer to client / customer as "finished" goods (Williams, 2003). Production chain is a way of collaboration among manufacturing enterprises to improve the response to changes in upstream and downstream production stages and to share lifecycle production data / knowledge. Successful production chains will be able to meet internal (business) and external (social) objectives with optimized performance in terms of cost, quality, time,

Hossam A. Gabbar (ed.), Modern Formal Methods and Applications, 37–46.

environmental issues, risk, etc. Production chains are formed with common objective to produce final product(s). For example, car manufacturing enterprises are linked to tire, glass, seats, mirrors, and lock manufacturing enterprises with the final objective to produce efficient car as a final product. It is part of the production chain design to find the strategic partner for productive collaboration.

Basically, production chains are virtual organizations that have lifecycle as composed of: concept stage, design, and operation stage. The operation design is usually starts during the concept and design stages of the underlying production chains and iteratively tuned till its maturity.

The design and execution of the operation of the underlying production chain is a complex process, which requires systematic way to integrate lifecycle aspects of process, production and management levels. In addition, it requires linking the upstream and downstream enterprises along with their hierarchical management and control schemes. To consider all aspects of production chain operation, formal representation framework is required to enable the definition of the different elements in the different hierarchical levels of the underlying production chain operation.

This research work proposes the use of formal methods to design the operation of production chains based on knowledge representation framework. The use of formal methods will facilitate the design and synthesis of production chain operation in different hierarchical levels and will provide robust base for automated engineering solutions to synthesize and manage the operation of production chains.

In the next section, operation framework will be described where operation model elements are explained for the different hierarchical operation levels. The third section describes the proposed formal method and knowledge representation approach, which is used to represent the operation of process level (i.e. micro) and production level (i.e. macro) of a case study PET bottle production chain.

3.2 Production Chain Operation Framework

The operation of system or process can be viewed in two main tracks: steady and transient operation. Steady operation is related to the operation during the steady state, which is based on the inventory control scheme. Transient operation deals with the operation during the changes in system state such as startup, shutdown, or fallback recovery. Operating a process (or system) implies satisfying safety, environmental, quality, management constraints and process conditions in all hierarchical control levels. For example, the operation of batch process might have one main objective to achieve the stated operation objectives such as production rate, production amount, and product quality, while satisfying safety constraints and process controls.

Production chain operation can be defined in hierarchical levels, as shown in figure 3-1. The operation of virtual organizations (i.e. VO's) includes tasks to supply material and resource as well as products between enterprises / organizations. Enterprise operation handles tasks among the different departments and units within an enterprise. Plant level operation handles tasks within one plant to produce the required product from the stated resources. Process level operation defines the tasks and controls to produce the expected products for the underlying process. Equipment level operation changes the status of the state descriptor of the underlying equipment.

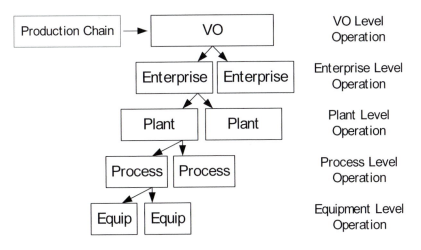

Figure 3-1. Hierarchical Levels of Production Chain Operation

3.2.1 Unit Operation Knowledge Model

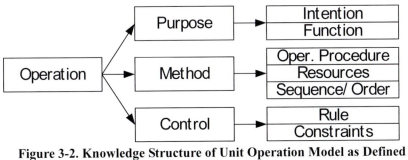

Figure 3-2. Knowledge Structure of Unit Operation Model as Defined within POOM

In order to build complex operation for large scale systems / processes, it is essential to define robust knowledge structure for unit operation, which will be used as building blocks for the complete operation of the underlying system / process. Plant/Process object oriented modeling methodology or POOM, showed how process model could be constructed in three dimensions: static, dynamic, and operation. Operation knowledge can be structured in view of the proposed operation model as expressed in POOM. In each of the hierarchical levels of production chain operation, operation knowledge can be structured as a collection of operation elements that includes: purpose, methods, and control, as shown in figure 3-2. The method includes actions, procedures used to achieve the intended function using set of resources in specific sequence. Rules and constraints are used to determine the control framework for the underlying operation. During the design and execution of such operation model additional knowledge will be accumulated such as the execution timing, which includes planned, scheduled, and actual time.

The above proposed operation model is mapped to other views within POOM, where function is linked to the dynamic view i.e. behavior model, while the resources include structural units and materials from the structure view. The method and control are mapped to the operation view. The intention and function are usually mapped to the objective and plan defined in higher level i.e. in the hierarchical production chain model.

UML proposed unified processes, which are best practices to realize target functions. Production chain model is constructed using building blocks of unit processes or UP, as shown in figure 3-3. Each UP will have input ports, output ports, and is described in three views: structure – SM, behavior – BM, and operation – OM. In such UP Model, "IS" is input stock UP and "MN" is manufacturing UP. Ports are used for energy, material, control, sensor, or information transfer.

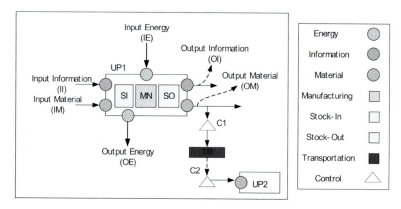

Figure 3-3. UP Model of POOM

The highest level of the whole production chain can be viewed as top level UP, which is composed of smaller UP's. For example, UP1 is composed of UP1.1 and UP1.2, etc. The operation model of each UP will be described in the form of OM of POOM.

In higher level within the production chain model i.e. enterprise UP, the enterprise OM can be described using the knowledge structure defined within POOM. For example, "MAKE" OM is used to define the process to manufacture product(s) and byproduct(s) from input raw materials, as shown in table 3-1.

Table 3-1. "MAKE" OM of the Enterprise UP, on the basis of POOM

OM1: MAKE **Function:** Make product from raw materials / intermediate products **Intention:** - Manufacture product(s) and byproduct(s) - Use raw material from input stock and produce products / byproducts into output stock - Fulfill production orders **Method:** - Plan required raw material as per required production rate/amount - Get raw materials from input stock - Operate production line with specified configuration - Stock products and byproducts in the output stock **Control:** - If raw material is not available, then replenish stock - If production line is busy, then wait or cancel operation - If output stock is full, then stop production or find extra stock space

Within the enterprise UP, there are smaller UP's such as stock UP, as shown in Figure 3-4. In such UP, process variables are defined for input material (i.e. IM-Flowrate for the input material flow rate), output material (i.e. OM-Cost for the unit cost of output material), and warehouse (i.e. St-OperCost for the unit operating cost). Two operations are defined for the stock UP: receive and issue. The "receive" operation is explained on the basis of POOM, as in table 3-1.

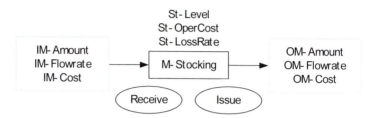

Figure 3-4. Material Stocking UP Model

Table 3-2. "RECEIVE" OM of the Stock UP, on the basis of POOM

OM2: RECEIVE
Function: Receive Material from output material port into warehouse
Intention:
- Add material to stock
- Transfer material from input port into warehouse
- Increase stock
- Replenish stock
- Return material to stock

Method:
- When material is available in input port, Open material input port
- Check material
- Decide suitable warehouse location
- Take material from input port into the decided warehouse location
- Update stock level of that material

Control:
- If material is not suitable, then reject material
- If no space available, then reject material
- If stock exceeds maximum stock level, then reject material

Similarly, other operations can be defined, such as stock take (reconciliation). Similarly, operation such as "COOLING" from batch process can be represented using the same operation knowledge structure model, as shown in table 3-4.

Table 3-3. "COOLING" OM for batch process

OM2: COOLING
Function: Cooling of structure equipment Receive Material from output material port into warehouse
Intention:
- Add material to stock

- Transfer material from input port into warehouse
- Increase stock
- Replenish stock
- Return material to stock

Method:
- When material is available in input port, Open material input port
- Check material
- Decide suitable warehouse location
- Take material from input port into the decided warehouse location
- Update stock level of that material

Control:
- If material is not suitable, then reject material
- If no space available, then reject material

From the above two examples, which are taken from two levels of the hierarchical production chain model, there are two major challenges: finding a formal method for systematic representation of the operation model elements, and linking the operation model elements in different hierarchical levels, such as the "RECEIVE" OM in stock UP and "MAKE" OM in the enterprise UP.

The following section will explain proposed solution to address these two issues, based on formal methods and knowledge representation techniques.

3.3 Formal Representation of OM

In order to achieve efficient production chain operation, it is essential to provide suitable representation mechanism for all operation model elements, which is suitable for all hierarchical operation levels.

The following section explains the proposed model formalization method and its use to construct the underline production chain model on the basis of operational design concept. The third section describes the proposed control layer and its mechanism to support steady operation of production chain. The fourth section explains the utilization of the control layer to synthesize operating procedures for production chain.

Table 3-4 shows examples of generic inventory control rules that can be applied to any structure class marked with inventory flag as "Yes".

Table 3-4. Examples of Generic Inventory & Process Constraints

IF INVENTORY LEVEL = MAX THEN CLOSE UPSTREAM MATERIAL
CONTROL
IF INVENTORY LEVEL = MIN THEN OPEN UPSTREAM MATERIAL
CONTROL
IF PROCESS VARIABLE = VALUE THEN OPEN UPSTREAM
PRODUCT CONTROL

Inventory rules are used to control the upstream material control devices, while process constraints are used to control downstream or upstream control devices. If conflict occurs in one level (i.e. Open C1 and Close C1), the proposed control layer will resolve such conflict by consulting control layers of upper control levels.

3.4 Case Study Production Chain

For example, table 3-5 shows operating procedures, which are represented using EFL, to move produced product from UP1 to UP2. In this case topology analyzer will define the associated topology area boundaries, which are C16, C17, and C19.

Table 3-5. EFL for Operating Procedures to Move Produced Product from UP1 to UP2

OPEN C18
CLOSE C19
CLOSE C16
CLOSE C17

Figure.3-5 shows the visualization of the above operation, where red circle is used when the control device is closed, while green circle is used to show that the control device is opened.

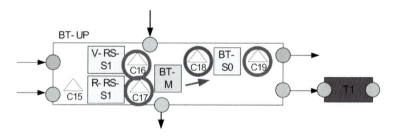

Figure 3-5. Visualization of the Operation "Move Bottle FROM UP1 TO UP2"

3.5 Conclusions

Typically, plant operation is executed in hierarchical manner where process level operation is integrated with production level operation to ensure smooth integration of the different hierarchical control levels.

In this chapter, POOM modeling methodology is used to construct plant model as building blocks called UP's. Each UP is represented in three views: static, dynamic, and operation. POOM is used to construct operating procedures in the different hierarchical levels. EFL or engineering formal language is proposed to represent operating procedures in the different hierarchical levels, which has been implemented within computer-aided modeling environment called CAPE-ModE. In such environment, operating procedures are visualized to show the hierarchical operating procedures, such as "OPEN" or "CLOSE" of actionable structural units.

The proposed solution is effective for automatic synthesis of operating procedures (SOP) in hierarchical levels of production plants. It will facilitate the job of operators to understand the different hierarchical operation and take right decision in less time.

3.6 References

ANSI/ISA-95 (2000). Enterprise-Control System Integration Part 1 & Part 2. American National Standard.

Aoyama, A., Naka, Y., Shimizu, A., Hamada, K., Kameda, K., Kagiyama, T., Matsumoto, I., and Tsujikawa, Y. (2003). GPLS: A modeling and simulation system for product lifecycle design. 4th International Conference on Foundations of Computer-Aided Process Operations, January, 2003, Florida, USA.

Gabbar, H.A. and Naka, Y. (2003). Control Mechanism for Production Chain Operation. 12th International Conference on Computer Theory and Applications (ICCTA'2003 – IEEE), 26-Aug-2003, Alexandria, Egypt, P28.

Gabbar, H.A., Chung, P.W.H., Suzuki, K., and Shimada, Y. (2000). Utilization of unified modeling language (UML) to represent the artifacts of the plant design model. Proceedings of "PSE Asia 2000" International Symposium on Design, Operation and Control of Next Generation Chemical Plants, PS54, 387-392, Kyoto-Japan.

Japan Chemical Innovation Institute (JCII) (2001). Development of Plastic Production Chain With Recycling", Project Report, Tokyo Institute of Technology, Yokohama, Japan, Mar-2002.

Johnson, M. (2003). Transforming B2B exchange into collaborative trading communities. MaterialWorld technologysolutions, Miami Beach Convention Center, 17-19 March, 2003, www.techexchange.com/thelibrary/exchange_to_collab.html.

Lakhal, S., Martel, A., Kettani, O., Oral, M. (2001). On the optimization of supply chain networking decisions. European Journal of Operational Research, Vol. 129, 259-270, 2001.

Lu, M. L., Batres, R., Li, H. S., and Naka, Y. (1997). A G2 based MDOOM testbed for concurrent process engineering. Computers & Chemical Engineering, Vol. 21, Suppl., pp. S11-S16.

Lu, M. L., Naka, Y., Shibao, K., Wang, X. Z., and McGreavy, C. (1995). A multi-dimensional object-oriented model for chemical engineering. In Concurrent Engineering, A Global Perspective, Virginia, Aug. 1995, Concurrent Technology Corporation, USA, pp. 21-29.

Naka, Y., Aoyama, A., Shimizu, A., Hamada, K., Kameda, K., Kagiyama, T., Matsumoto, I., Tsujikawa, Y., Ranajit, C., and Gabbar, H.A. (2002). Development of plastic production chain with recycling. Technical Report of Japan Millennium Project, Yokohama, Japan.

Naka, Y., Batres, R., and Fuchino, T. (1999). Operational design and its benefits in real-time use, Foundations of computer aided process operations (1999), ISBN 0-8169-0776-5, pp: 570.

Naka, Y., Hirao, M., Shimizu, Y., Muraki, M., and Kondo, Y. (2000). Technological information infrastructure for product lifecycle engineering. Computers and Chemical Engineering, Vol. 24, 665-670, 2000.

Naka, Y. and McGreavy, C. (1994). Modular approach for startup operational procedures of chemical plant. Proceedings of PSE'94, 1007-1013.

Supply Chain Council. SCOR Model. www.supply-chain.org.

Eric Williams (2003). Forecasting material and economic flows in the global production chain for silicon. Technological Forecasting & Social Change, 70 (2003) 341–357.

Zhou, Z., Cheng, S., Hua, B. (2000). Supply chain optimization of continuous process industries with sustainability considerations, Computers and Chemical Engineering, Vol. 24, 1151-1158, 2000.

4 Formalizing Waste Management

Author
Pohjola V.J.
University of Oulu
Finland

Summary
PSSP ontology is introduced as a basis for holistic worldview and as an approach to formalize and systematize any domain. PSSP ontology defines Event and Medium as two primitive kinds of objects that more complex objects of Process or Product kind are composed of. In practice anything is a process. All objects have four predefined properties: Purpose, Structure, State and Performance. The result is a unified representation of reality. PSSP approach is applied to formalize and systematize waste management. This is a challenging task, because the domain has extremely fuzzy boundaries and the waste management decisions must be based on integrating knowledge of technical, social and ethical issues.

4.1 Introduction

Waste management is a domain, which through centuries has stubbornly resisted becoming systematized. Although pervading all human civilization, the domain lacks a holistic methodology or theory of waste management. This is primarily due to the lack of commonly accepted universal definition of waste. The major challenge for systematization efforts thus resides in the confusion concerning what is it we call waste: How to manage something you cannot take a hold?

Because of its pervading nature the boundaries of the domain of waste management are fuzzy. It is not possible to say strictly which issues form the substance of the domain and which ones are irrelevant. Thus the starting point of systematization needs to be an ontological one and the ontology to be adequate instead of domain ontology.

Hossam A. Gabbar (ed.), Modern Formal Methods and Applications, 47–82.
© 2006 *Springer. Printed in the Netherlands.*

This paper introduces the PSSP ontology as the formal approach applied for defining waste and, on that basis, for systematizing waste management. The adequate PSSP ontology provides a system, which pre-defines the set of universals: Purpose, Structure, State and Performance, as the necessary and sufficient set of properties of objects of all kinds. Thus the primitive objects of both Event and Medium kind manifest their way of being through these four properties. Also more complex objects of Process and Product kind, which are aggregates of event-medium composites, are characterized by these properties.

It is shown in this paper that the PSSP ontology gives a totally new insight into the domain of waste management. A new set of definitions of waste emerges from the holistic perspective opened. The attributes Purpose and Performance have the central role in these definitions. Ownership as an abstract object of Relation type plays a central role both in defining waste and in systematizing waste related activities.

4.2 The formal method

Ontological approach does not take as its starting point the current usage of expressions of natural language, which are vague and understood differently by different people. Neither does it accept as its starting point the multitude of practices in a domain, as standardization efforts normally do. The ontological starting point is our a priori knowledge of reality as it is in itself, that is, what we know about the *kinds of objects* there can be and the *kinds of properties* they can have. It is against such a framework of knowledge of *universals* that our domain knowledge of *particulars*, based on perception and reasoning, is *identified* and *specified*. This is what Lowe has named a four-category ontology [1], and can be depicted as in Figure.4-1.

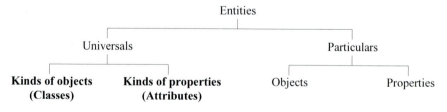

Figure 4-1. The 'four-category' ontology

The ontological approach taken to waste management does not suggest that the terms 'waste' and 'waste management', or the current waste management practices, were useless and should be replaced by something else, but only that they can be, and should be, re-interpreted and re-engineered in a new conceptual framework, which is holistic as opposed to being restrained by the narrow scope and bias of tradition.

4.3 PSSP ontology

The formal method to be applied to identifying and specifying waste management is built upon the PSSP ontology [2, 3, 4]. The name of the ontology is an acronym of the four universal properties or *attributes* that objects can have: Purpose, Structure, State and Performance. The basic commitment of PSSP ontology is that this set of attributes is the necessary and sufficient set of attributes of all objects. It should be easy to apprehend that this commitment leads to a highly unified worldview, which has crucial consequences for our thinking, like becoming more holistic and more creative.

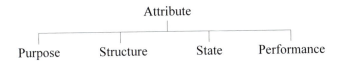

Figure 4-2. The four universal properties

What we perceive are *differences* and *similarities* in properties of objects. While what we perceive and reason is subjective, the fact itself that objects can have similarities and differences is independent of human thinking. To be similar or different implies that objects must have properties. Attributes are our a priori knowledge of reality ('as it is in itself') in the sense that no empirical knowledge, that is, knowledge based on perception and reasoning, can be articulated without first knowing what properties there can be. Purpose, Structure, State and Performance are through which an object *manifests* its way of being, independent of, and to be distinguished from, whether or how truthfully human is capable of characterizing an object by *specifying* the four types of properties.

On the basis of what we know by perception and reasoning about a particular object's properties we aim to *identify* the object as belonging to some object *class*. Objects can be similar or different with respect to one or several properties and to varying degrees. That is why classifications (taxonomies) based on empirical knowledge can be almost whatever. There are, however, some similarities and differences, which are relatively easy to accept as fundamental and of which we can intuitively have knowledge of a priori type. This is ontological knowledge, which forms the topmost layer of most of the proposed hierarchies representing our knowledge of what can be.

4.4 The universal properties

An object manifests its particular way of being by having particular properties, which are instances of the four universal properties, Purpose, Structure, State and Performance. An object's way of being is a whole, not a collection of individual properties. This is to say that the four properties are deeply related to each other. An object's potential to function, which becomes manifest as its behavior (*state*) in the object's coexistence with other objects, is embedded in the object's *structure*. The way an object behaves in particular situations is its *performance* when set against how the object should behave in all these situations in order to justify its existence, that is, against its *purpose*.

4.4.1 Purpose

Purpose is an object's *claim of justification to exist* (currently, in the future, or in the past). Hence, all objects manifest their purpose by existing (actually coexisting) and surviving. For an object to survive is to adapt its behavior to face the challenges of the coexistence. Capability of adaptation arises from an object's *functionality* (potential to function), which is hidden in an object's structure. An object has a given structure and, consequently, a given functionality hidden in that structure, just because there is a claim of justification for that structure. Thus, an object's purpose becomes manifest in its structure and in how the object behaves in particular situations, that is, in particular relations to other objects. Fulfilling what is claimed is by which an object justifies its existence and survival. Fulfillment can be of varying degree and vary with time, which entails that an object once justified to exist may not survive in other situations and would then be doomed to lose its identity.

Purpose is a universal property no matter whether it is possible for human to figure out what might be the purpose of a particular object. This means, in particular, that in the PSSP ontology, purpose is a property of both natural objects and artificial objects (artifacts). Aiming to justify the existence of a particular *artifact* is *specifying*, that is deciding upon, its purpose by the owner (designer, builder or user) of the artifact. On the other hand, whatsoever may be the purpose of a particular *natural* object, whose identity and existence are not dependent of human, is not to be decided, and in many cases not even hypothesized, by human. Indeed, whether a particular purpose of an object *can* be known to human can be used as an indicative guide for distinguishing artificial objects from natural objects. The borderline between natural objects and artifacts is, however, fuzzy. Sometimes an object can be identified as an artifact by some of its structural features (like type of material or shape of boundary) but the original purpose may remain unknown, because the links to the production or usage processes are difficult or impossible to trace (take an archeological finding as an example). Nevertheless, in this case, the purpose *could* be known if sufficient knowledge were available. Human being is a good exemplar of an object, whose purpose as a human being tends to remain a mystery to human being himself, and, on that basis, human being should be viewed as a natural object. However, during his life, each human being is assigned a multitude of purposes in various roles as a member of society, which supports viewing human being as an artifact. For instance, an employee fired loses his identity in the role he had in his job. In the case of not having other roles and the associated assigned purposes, he would find himself perplexed by his purpose as a natural object.

An object manifests its purpose under the constraints arising from its relation to other objects. The constraints have a role of an object's *performance* criteria. The extent of fulfillment of the claim of an object's justification to exist is the object's performance. If sufficient, the existence of an object in the given relation (coexistence) is justified. Thus, purpose and performance are inseparable.

Characterizing reality in the PSSP format is using the PSSP ontology as a *language*. Human specifies and informs about the purpose of a particular artifact by assigning a value for the Purpose attribute of the PSSP object *representing* the artifact. In design, where the artifact under design does not exist yet, the value is assigned as a requirement or expectation concerning the functionality (potential to function) of the artifact. The value of the Performance attribute is an assessment of the artifact's performance against the performance criteria listed and weighted under the Purpose attribute.

4.4.2 Structure

Structure is a hierarchy of *relations*. An object can manifest its structure as a set of sub-objects and a set of causal, spatial and temporal relationships between state variables of these sub-objects. Each of the sub-objects can do the same, and so on infinitely. This applies to both natural and artificial objects. It is easy to sympathize Korzybski's stand, that whatever human says about structure it is more [5]. On the other hand, depending on the purpose of a particular description, only certain aspects of an object's structural information is relevant.

In PSSP ontology the structural hierarchy is viewed to expand in two dimensions, topological (or fractal) and unit-structural. In the former, an object disaggregates into sub-objects such that the object and the sub-objects are instances of the same class. The mutually linked sub-objects (*nodes*) form the *topological* structure. In the latter dimension a node decomposes to sub-objects (*parts*), which are instances of other classes than the node and, as linked to each other, form the *unit* structure of the node. Structural specification of an object in the two dimensions results in a powerful description, which can be continued to any level of detail.

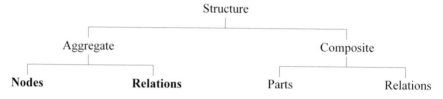

Figure 4-3. Expanding of structural hierarchy in two dimensions

An object's structure embeds all the information about object's functionality. If an artifact has been designed ignoring some structural parts and links (like a customer's usage process), it may not have proper functionality (may not behave as required in that usage process) resulting in poor performance. If an artifact breaks, some of its structural parts and links assume other than intended values, and some of the intended functionality is lost. This shows in the artifact's behavior and performance.

4.4.3 State

State is how an object's functionality becomes manifest in particular situations of coexistence. State may appear to human as static or dynamic. Quoting Bohm [6]: 'Whenever one *thinks* of anything, it seems to be apprehended either as static, or as a series of static images. Yet, in the actual experience of movement, one *senses* an unbroken, undivided process of flow…'. Dynamic state is often called *behavior*. Dynamic state may appear as transient like an object's transition from one stationary state to another, or as pseudo-stationary like an object's continuous fluctuation around equilibrium or some other desired steady state.

State

Dynamic (Behavior)   Static

Figure 4-4. Semantic relationship between state and behavior.

In PSSP ontology, state refers to *temporal distribution* of the state variables of an object, where state variables are those distributed amongst the structural relationships in the two dimensions (topological and unit-structural) throughout the object's structural hierarchy. Description of a particular state, based on perception and/or reasoning, can be done in any appropriate way, numerically, graphically or verbally, in the form of empirical time-series data, mathematical function, results of numerical simulation, as a snap-shot at certain location on the time axis, etc.

Complex objects usually have a structure, where the state variables distributed throughout the structural hierarchy are also in a hierarchical relationship to each other such that some states *control* some other states. This is of course the principal way for functionality to become enriched both among natural and artificial objects.

4.4.4 Performance

Performance is by which an object manifests how it manages to fulfill its justification to exist and to survive. Because the challenges of coexistence are usually manifold, survival typically implies readiness to many types of behavior. An object may need a structure, which provides functionality for effective transitions from the current to more beneficial 'operating conditions' or for maintaining the current (pseudo-stationary) state distribution in face of disturbances. It is here that various control structures and strategies play a central role.

Performance can be viewed as a measure of how good an object is for its purpose. An object capable of self-control and self-organization assesses its own performance against the built-in performance criteria. When an external agent assesses the performance of an object, the assessment must be based on performance criteria set by the agent. In particular, assessment of the performance of a natural object is not possible for human unless a hypothetical purpose is specified first. Of course an external agent's opinion of an object's right to exist is as justified as is the external specification of its purpose.

4.5 Central objects

In PSSP ontology, the top-level object class Entity divides into two sub-classes *Event* and *Medium*. The division is intuitively appealing due to the fundamental difference between the generic structure of events and the generic structure of media. Consequently, there are generic level differences also in the other properties of the two types of *primitive* objects.

Events are known to be intangible, advancing in time and becoming manifest only via the medium in which they occur. Medium, thus, holds information about events. Information about events that *can occur* in a medium is embedded in the structure of the medium. Information about events *occurring* in a medium is manifested by the medium's state (behavior). Obviously events cannot exist independently but always associated with some medium. Similarly, if there is no event occurring in a medium, there is always at least a potential in any medium for some event to occur. Thus, events and media coexist as composite objects.

Event-Medium composites can be regarded as the operative primitives of reality. They are the building blocks of yet larger composites or aggregates called *processes* and *products*. Process and Product are two object classes with high structural similarity but differing by purpose, state and performance. Practically any object can be identified either as a process or a product. This applies to waste and waste management as well, as will become apparent.

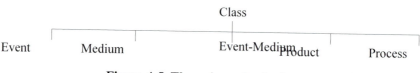

Figure 4-5. The universal sub-classes.

4.5.1 Event

The division of reality into events and media is not only ontologically fundamental but also practical. Events are a pervading type of objects, including spontaneous physico-chemical phenomena and human mental and physical activities and are mostly easy to recognize. Event is distinguished from medium by its inner structure, or *unit structure*, to apply the PSSP usage. While event can disaggregate into sub-events and these further into sub-sub-events, etc., each of these has a unique unit structure composed of the predefined parts, *cause*, *effect* and *precondition*, and *causality* linking the parts.

Causality is rather the name of the type of relation associated with event, linking cause to effect and preconditions, than commitment to an idea that all causal relations were deterministic or could be specified. Causality is the mechanism by which an event advances. Sometimes, like for some physico-chemical phenomena advancing in matter, it is possible to construct a deterministic theory (often based on probabilistic considerations) for specifying the causal relation and to predict the effect from cause and preconditions. Causality applies to human activities as well, but at least at an individual actor's level the mechanisms are unknown. Fortunately human, as opposed to inanimate matter, can explain his doings. Thus a human actor may be asked to report how he is going to act, or why the activity advanced as it did. This *rationale* can be viewed to correspond to causal relation and forms an invaluable type of knowledge concerning human activities.

The *state* of an event is its *rate* and *extent* of advancing. These state variables are to be understood as distributed along time axis. Rate is linked to the effect and extent to the cause. While rate can be regarded as the principal state variable, extent applies when an event pursues a goal. Goal is usually a desired state of the medium. Rate can be constant, in which case an event advances steadily, while extent approaches the value corresponding to the goal. Otherwise event is in a dynamic (accelerating or decelerating) state. When an event does not advance, its rate is zero. This state may be associated either with an event, which does not exist, or with an event which does not exist yet but whose potential to be born is in the structure of a medium. An event may be apparently non-advancing when idling after the goal has been reached, which is the state fluctuating around the desired state.

Because to exist for events is to *advance*, the general-level *purpose* can be expressed as follows. For physico-chemical phenomena the purpose is to advance under the constraints posed by Nature like those, which have been formalized as entropy principle and Hamilton's principle, and which we use to explain these natural events. The constraints apply no matter whether the medium is natural or artificial. The purpose of human activity is to advance under the complex of natural constraints concerning body and the mental constraints concerning mind.

Events cannot be seen and thus not depicted figuratively. In PSSP formalism an event is *referred* to by a horizontal oval as shown in Figure.4-6. The same notation is used when *characterizing* a particular event by specifying its attributes (Figure.4-6 (c)).

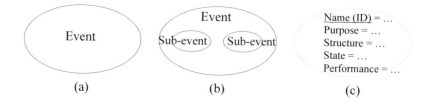

(a)   (b)   (c)

Figure 4-6. Graphical notation of event: (a) single event, (b) aggregate of two events, (c) template for characterizing an event.

4.5.2 Medium

The ontological division of objects into events and media makes it possible for human to identify an object to be a medium, simply whenever an object, as a constituent of an event-medium composite, does not identify as an event. Medium's *structure* is typically an aggregate of tangible and intangible sub-media, in which an intangible medium (energy, mind) becomes manifest via tangible medium (material, body). Individual aggregates can build up larger collectives like continuous matter as a population or aggregate of molecules, or organization as a group of human beings. A collective medium's functionality resides in the relations both between and inside the lower level aggregates.

A medium's functionality remains latent and its *state* only spatially distributed, if no event occurs in it. Consider an artifact (product), made of solid material, lying on a vendor's shelf. It most certainly has some functionality as a product, because it was designed and manufactured to have one. However, the material (medium) does not manifest any other behavior than its stagnancy in terms of spatial distributions of state variables like thermodynamic state variables temperature, density and weight fractions of sub-materials. Apart from its functionality as an artifact, originating from human, there is in the material a potential for some spontaneous phenomena to occur. If, for instance, the material is heat conducting, there is potential for heat conduction to take place. Heat conductivity is one of the material's inherent functionalities embedded in its deeper structure. The deeper structure, of course, refers to macroscopic material as an aggregate of molecules or atoms.

Because medium and event are inseparable, it can be assumed that the general-level *purpose* of medium has to do with potential events. One is thus tempted to propose that at least one, if not the sole, justification for medium to exist is to make advancing of events possible. Medium provides a potential for an event to occur by having a given structure and by being in a given state. Besides providing the potential for heat conduction to occur, a heat conducting material provides the cause for heat conduction to advance, if there is a temperature gradient in the material. As another example, consider a mixture of two components as a medium providing the potential for the two components to react chemically with each other. Besides providing the potential by having the given structure, the medium provides the phenomenon with the cause and preconditions, by having the components in proportions not corresponding to chemical equilibrium (the minimum of the Gibb's energy) and by having a temperature level making the molecular level collisions sufficiently energetic for the chemical reaction to advance.

The graphical notation of Medium is fully analogous to that of Event as shown in Fig.4-7.

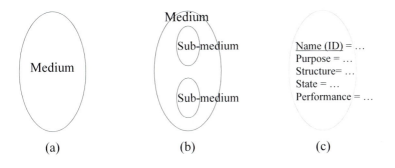

(a)   (b)   (c)

Figure 4-7. Graphical notation of medium: (a) single medium, (b) aggregate of two media, (c) template for characterizing a medium.

4.5.3 Event-Medium composite

Event-medium composites are, as the name implies, objects with a generic *unit structure* composed of an event and a medium interlinked by a relation of causal type. The properties of event and medium can be taken as two projections of the properties of the composite, viewing them from different angles and mapping into each other by the causal relation as depicted in Figure.4-8.

The rate at which an event advances at any moment is associated with its effect, which is a transition of the medium from a state to another. The cause for an event to advance is associated with a deviation of the medium's state from a desired state. Preconditions for an event to advance come from the medium's energy level. The causal relation, that is, how the rate of an event depends on its cause and preconditions is thus the same as how the change of the medium's state depends on its deviation from a desired state and on the medium's energy level.

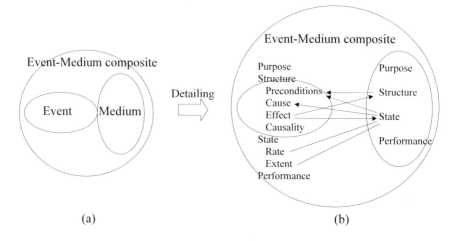

(a)   (b)

Figure 4-8. (a) Graphical notation of event-medium composite. (b) Ownership relation as a network of mutual monitoring/manipulation links.

Reality as a whole can be viewed as a single event-medium composite. In this *holistic* worldview both event and medium can be conceived of as aggregates of sub-events and sub-media, respectively, and these sub-objects as aggregates of even lower level sub-objects and so on. At any hierarchical level certain events and certain media are interlinked building up lower-level event-medium composites. Not any event can link to any medium. Which events *can* link to a given medium is dictated by the medium's structure, which is where the medium's potential to function is embedded. This potential has also been referred to as *ownership* relation [4,7]. By ownership one means the right or responsibility of a medium (or an event-medium composite) to manipulate its properties by providing an event the cause and preconditions to advance (see Figure.4-8.).

Several sub-events can occur in a single medium. Because information about a particular sub-event is obtained indirectly by making observations concerning the structure and the state of the medium, the interpretation of observations may be difficult when several sub-events are occurring simultaneously in a single medium, as is often the case.

A single event can occur simultaneously in several sub-media. The ownership relation is then between the event and the medium as a whole, that is, the aggregate of the sub-media. It is the medium as a whole, which provides the cause and experiences the effect of the event. This applies even when one of the sub-media is conventionally viewed as 'an actor' and another as 'a target' or 'a patient'. Such a view arises from the subject-verb-object structure of modern natural languages and has had a detrimental effect on our thinking [5, 6, 8]. The formal language built on PSSP ontology implements exactly what a language, which is in harmony with reality, should do: the central role should be given to the verb rather than to the noun. Quoting Bohm [6]: instead of saying 'An observer is looking at on object', we can more appropriately say 'Observation is going on in an undivided movement involving those abstractions customarily called 'the human being' and 'the object he is looking at'. In PSSP terms, the undivided movement is the top-level event-medium composite involving an observation event advancing in a medium, which is an aggregate of 'an observer' and 'a target'. Observation can occur only when the two sub-media make a whole. In other words, there is no ownership relation involved, and thus no manipulation of the properties of the medium (or the event-medium composite) can occur, if the two sub-media do not make a whole.

Figure.4-9 depicts three situations for characterizing the ownership relation: (a) full ownership in a truly homogeneous medium, like freedom of material or human mind to spontaneously reorganize, (b) socially controlled ownership in a pseudo-homogeneous medium, like restricted freedom to manipulate itself of an aggregate of humans, or an aggregate of humans and artifacts, and (c) indirect ownership across a boundary in heterogeneous medium, like interaction between interior and exterior of a process.

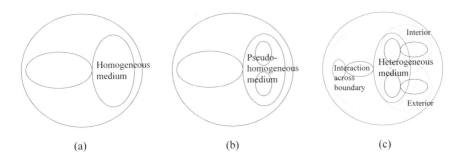

(a)   (b)   (c)

Figure 4-9. (a) Event occurring in a single medium. (b) Event occurring simultaneously in two sub-media taken as a single (pseudo-homogeneous) whole. (c) Event (interaction across boundary) occurring in a heterogeneous medium.

   Birth, survival and death of a particular event-medium composite follow
from reorganization of medium. For illustrating this statement, consider a
person reading a newspaper. In order that the reading activity can take place the
person needs to *have* a newspaper. A medium capable of manipulating its
properties by providing the reading activity (information transfer) with a cause
and preconditions to advance should be borne first. We may say ignoring the
surrounding air and a proper lighting that a person and a newspaper need to
own each other to make up an aggregate medium having that functionality. We
may also say that after the reading activity (information transfer) has reached
its desired state, the ownership, and thus the existence of the aggregate
medium, is no more necessary for that purpose.

   Continuing the previous example, the primary cause for a person to own a
newspaper is his *being aware* that there are newspapers equipped with
probably attractive information, while the primary cause for a newspaper to
own a reader is its being *meant* to be read and to have its information content
sufficiently attractive. In the realm of inanimate objects, random movement
rather than conscious search is the mechanism for objects to find each other.
After the ownership relation has formed between a newspaper and its reader,
the *cause* for a reading activity *within the aggregate medium* (like that in
Figure.4-9 (b)) comes from the medium's spatially unevenly distributed
information content, or a lack of information on the reader's side. Both the
level of intelligence of the reader and the legibility and comprehensibility of
the text are included in the *preconditions* of reading. An obvious *effect* of
reading is that the reader becomes more informed. Information flow differs
from material and energy flows in the fundamental respect that information
output does not decrease a medium's information content. Thus reading has no
effect on a newspaper's information content. However, as a part of the
aggregate the newspaper ultimately loses its primary functionality (although it
may still retain other useful functionalities). This happens when there is no
more cause and/or preconditions for reading left in the aggregate medium.
When detached from the aggregate the newspaper has, of course, all its original
functionality left to utilize by another reader.

Figure 4-10. Life cycle of an event-medium composite.

A newspaper and especially its information content remain intact under reading, because information flows in one direction only and does not imply simultaneous material transfer. The electromagnetic radiation (light) necessary for carrying the information is not generated but only reflected by the paper-ink medium and does not alter the medium's structure or state. From the effect's perspective it may be tempting to say that the reader monitors the text, and the text manipulates the reader. A more appropriate way is to say that reading (information transfer from newspaper to reader) advances in the aggregate medium, which is the manifestation of the medium's right to manipulate itself for moving it towards the desired state. Whenever there is actual material or energy transfer involved between an observer and the object being observed, also the latter experiences an effect. Maybe quantum measurement and the associated Heisenberg Uncertainty Principle is the best-known example. When an object being observed is human, the 'observer effect' is known to be present. Even prediction of human behavior, when made public, is known to have an effect on the actual behavior.

4.5.4 Process

Reality as a whole, if viewed as a single event-medium composite, has no bounds. On the other hand, if there is a *boundary*, there obviously should be something beyond the boundary and possibly some *interaction* across the boundary. If this view is taken, reality as a whole is rather a *process*, that is, an aggregate of what remain on both sides of the boundary, together with the boundary itself and the interaction across it.

Boundary is a peculiar object (see e.g. Smith and Varzi [9]). Boundary can be viewed as an intangible medium and as such perceptible only via the media on both sides. Some boundaries are just mental constructs and thus related to the way human reasons about reality. On the other hand, most if not all event-medium composites at lower levels of object hierarchy coexist as aggregates, which manifest their structure and obtain their identity via boundary. These aggregates are either *processes* or *products* having the generic unit structure composed of *boundary*, *interior*, *exterior* and *interaction*. As an illustration, consider an artifact (product), made of solid material. It manifests its structure via the solid-air interface, which is the boundary to be perceived by seeing or touching the solid material on the other side. On the other hand, the boundary of a project (which is a process) has no shape and cannot be perceived but only reasoned as a label specifying which objects there are in the interior and which belong to the exterior.

Interaction is an event of specific type. It is a material, energy or information transfer phenomenon or activity occurring across a boundary between the media on both sides. Thus the medium necessary for providing a potential for interaction to occur is an aggregate of the internal and the external media and the boundary. The cause for transfer phenomenon or activity is a deviation from a desired state of the aggregate. Usually this means that there is a lack of material, energy or information on either side. Boundary's contribution for the causal relation is in controlling the transfer rate.

Boundary obtains its properties like shape and permeability from the media on both sides. The shape of the solid-air interface of a solid piece of material comes from the spatial distribution of the solid material. Boundary may be selective meaning that the transfer rates of different material, energy or information types differ from each other. That a solid piece of material can be touched stems from the fact that the air on one side of the boundary is fully permeable to hand while the solid material on the other side is fully impermeable. That a piece of metal feels different from a piece of wood on touching comes from various properties of the solid material like its surface pattern and heat capacity and the resulting differences in energy transfer across the boundary. If a piece of material deforms on touching, we may say that its boundary is permeable to mechanical energy. What sort of information is transferred, and at what rate, across a project boundary is a matter of agreement made between the human resources in the project's interior and those in the exterior.

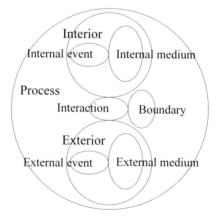

Figure 4-11. Process viewed as an aggregate of three event-medium composites: Interior, Exterior and Interaction-Boundary.

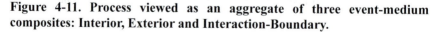

4.5.5 Product

At the level shown in Fig.5-1, process and product are structurally indistinguishable. The same graphical notation can be used for product. The difference is in how the structural parts, especially interaction, in reality manifest their way of being. During certain phases of its life cycle a product may manifest itself as a process. This is, when an artifact (a product of human origin) is under design or being used. In both situations a product needs to be understood to be a process, where an internal medium with specific built-in functionalities interacts with an external medium, which is the medium of a customer's usage process. When designed and implemented but not in use, an artifact is mostly passive and just waiting for an ownership, which would appreciate its specific functionalities and make the intended interaction active. In this passive state, like when a fresh newspaper is still in a mailbox, a product is just a medium equipped with a boundary, which gives it an identity as a product.

By the same token, during certain phases of its life cycle a process may manifest itself as a product. A process is the product (artifact) of a process design or embodiment project. Process refers then to documented information about the process to be built, or to the embodiment to be taken into use.

4.6 Application to waste management

4.6.1 What is waste?

'Waste', or its equivalent in other natural languages, is a word used to refer to objects *regarded as waste* by people in various situations in various cultures. Thus, wastes are linked to human assessment, and their existence to human civilization. The concept of waste has a meaning only in the context of artifacts, which are objects of human origin.

In this paper, a new set of definitions of waste is proposed. Based on that set a new perspective into what is called waste management is opened. A particular emphasis is on changing the focus more onto proper design and manufacturing of artifacts for preventing or minimizing waste *formation* from techniques of treatment of *existing* waste.

4.6.2 Traditional view

The following list [10] puts together a set of definitions of waste launched by a few major organizations:

EU: 'Waste shall mean any substance or object in the categories set out in Annex I which the holder discards or is required to discard.'
OECD: 'Wastes are materials other than radioactive materials intended for disposal, for reasons specified in this Table.'
UNEP: 'Wastes are substances or objects, which are disposed of or are intended to be disposed of or are required to be disposed of by the provisions of national law.'

Quoting earlier papers [8, 11] the common denominator in these definitions is that waste is something that the holder has disposed of/discarded or is going to dispose of/discard. Principally, both 'dispose' and 'discard' mean 'abandonment', perhaps 'disposal' is more putting it in a suitable place, while discard has the connotation of being useless or undesirable 'tossed aside'. It was assumed that the purpose behind the use of the expression 'discard' instead of 'disposal' by the EU Directive was to broaden its reach, and in the final destination of the discarded things. An interesting approach is to try to replace the term 'waste' with 'a thing that the holder discarded/intends to discard'. 'Minimizing the amount of things that the holder intends to discard' does not appear to embody the essence of waste prevention; it can be argued that it encourages re-use, recycling or recovery measures. The type of definition, 'a thing what its holder discarded' assumes that the waste is already there and the holder intends to dispose of it, while the principal meaning of waste minimization is to avoid waste production at the source. This waste definition thus fails to support the highest-ranking waste management option. The problem with the waste definitions listed above, is that they deal with existing waste. Such definitions seem to accept the fact that people/institutions throw things away, and therefore, existing legislation appears to be concerned with the 'what to do with it'-dilemma. This is understandable, as the main goal of European legislation on waste is the protection of public health and the environment. To conceptually describe waste is not the main purpose of these definitions. The label 'waste' does not necessarily mean that the thing is an ultimate waste; rather it means that it will be treated as waste. It appears that it is not feasible to come up with a comprehensive definition that unambiguously categorizes every discarded object as waste or not.

Quoting further the same source [loc.cit.], the goal of European waste-related legislation is to protect public health and the environment, and so far it has had a significant impact. However, given the lack of precision of the definition of waste in the European Community's Directive each Member State makes a different interpretation of the definition of waste with regard to specific materials, resulting in trade barriers and the impact of this upon the recycling industry is not to be underestimated. Under the present European definition of waste, recoverable material is seen more as a potential pollutant than as a potential raw material. As such, its movement between EU and non-OECD states falls under the restrictive conditions of EEC Council Regulations, if they are "hazardous". The essence of the legal definitions is that the owner does not want it; thus waste exists only where it is not wanted. Other definitions also explain why the owner does not want it. A proposed definition for waste is: "Either an output with ('a negative market') no economic value from an industrial system or any substance or object that has 'been used for its intended purpose' (or 'served its intended function') by the consumer and will not be reused". The second half of the definition suggests that the product was designed for one single purpose, and as soon as the purpose was fulfilled it turns to waste. It may still be functional, but it is no longer used nor re-used. It may also mean that the product lost its original properties and cannot fulfill its function anymore. On the other hand, the first half of the definition suggests that waste is a substance that no one ever wanted; it was created to be waste. The obvious question is why? Another problem with most of these definitions is that they do not suggest that creating waste is an unsustainable option. It seems acceptable to discard something no longer wanted, or to create something with no eventual long-term use at all. Yet, there are other types of wastes.

4.6.3   Waste as PSSP object

The traditional way of defining waste can be viewed as an inductive search of common denominators for objects regarded as wastes. The systematic search of a new definition of waste on PSSP basis comprises *identifying* the candidates of waste as belonging to one of the few fundamental PSSP classes: Event-Medium composite, Product or Process, and *characterizing* them in terms of the universal properties: Purpose, Structure, State and Performance. This top-down approach starts with Purpose as an object's justification of existence. The following proposition results immediately:

An *object*, which does not deserve being existent, *can* be labeled as waste.

But being waste is a *human* assessment. It is not in the hands of human to decide upon whether or not objects *in general* deserve being existent. In this respect there is, however, a difference between natural and artificial objects. The difference is in the nature of the *ownership relation*. The ownership between a human being and a natural object can be thought of being spontaneous and shared among all human beings, while between human beings (individuals and organizations) and artifacts it is given, taken, or traded, and private. For instance, people, like all creatures, can freely breath the surrounding air, that is, to manipulate its state in terms of the state variables like oxygen, carbon dioxide and water content. But no one can privately own all the air. Air can be used as a raw material for making artificial products like, say, liquefied air, which can be privately owned. It is *possible* that such a product appears to be of no use and not to deserve being existent. But it is air in the role of an artifact only, which can be labeled as waste. Based on the difference in the nature of ownership relation a more narrow proposition results (see Fig.12(b)):

An *artifact*, which does not deserve being existent, *can* be labeled as waste.

There still remains an ambiguity due to the fuzziness of the borderline between natural objects and artifacts, which cannot be fully eliminated, and a possibility for ethical problems to arise from the ambiguity, which needs to be recognized. As discussed earlier, *purpose* is the property, which can be used to draw a line between natural and artificial objects. An object's belonging to either of these sub-classes may be tentatively tested by whether or not it is *possible* for human to *know* its specific purpose.

With an artifact it is the matter (right or responsibility) of human in the role of a *private* owner (designer, producer or user) to *decide* upon its purpose. To decide is to manipulate. Thus, the one who decides that the purpose of an object is that and so can be said to know that the purpose is that and so. People, who are not private owners of a particular artifact, may recognize it as an artifact by its structure or state (or behavior), and on that basis know that it *must* have a purpose assigned to it by its private owner, although they do not necessarily know what it is. For instance, parts of technical artifacts have specific purposes, assigned to them by their designers. Typically these are not known to ordinary people, not even in the role of a private owner of such parts. In many cases people even do not care to know the purpose, if a part behaves as it should and thus provides the device, that it is a part of, with acceptable performance. Yet, by knowing that a part behaves properly, they know that there must be *a* specific purpose (functional requirement) assigned to the part.

To summarize, an artifact's owner has a right to manipulate the artifact including decision upon its purpose. Under certain constraints certain types of natural objects can become privately owned and thus be assigned a purpose. *Deciding* upon an artifact's purpose (or assigning a purpose) is to *manipulate* an artifact in the same sense as changing an artifact's structure is to manipulate an artifact. Manipulation is usually intentional and built on some knowledge. *Characterizing* an artifact's purpose does not imply ownership. In order to characterize, a person *monitors* (perceives and reasons about) an artifact, or becomes informed by the owner. In most cases monitoring does not involve manipulation. The information internalized, that is, extracted or received and personally interpreted, becomes the person's subjective knowledge of the purpose. That knowledge is the person's capability of characterizing the artifact (see Fig.12(a)). So, people either know or do not know what is the purpose of an artifact, and in the case of not knowing, they know that they possibly *could* know if informed properly. Also people know that they possibly *cannot* know what is the purpose of a natural object. People know that it is possible for human only to hypothesize, not to know and least to decide, what the specific purpose of a natural object might be. Thus, it is obvious that waste cannot be defined without reference to *knowledge*.

When knowledge, which leads to characterizing an artifact as a candidate of waste, is *externalized* in PSSP format, it either has its Purpose specified as 'None' or its Performance specified as 'Unacceptable'. When an artifact acceptably has no purpose, its performance does not matter. An artifact with unacceptable performance always has a well-specified purpose.

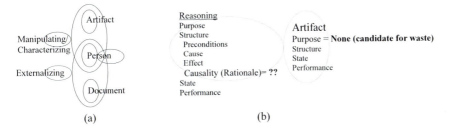

(a)   (b)

Figure 4-12. (a) Externalizing private knowledge. (b) Information content of a document

An artifact becomes a candidate of waste when manipulated by its owner and characterized as having no justification to survive. The crucial question is, what should be required of the knowledge that the acts of manipulation and characterization are based on, in order for the acts to be justifiable leading to declaring an artifact as waste. This question extends the task of defining waste one level up (to metalevel) and calls for metaknowledge, that is knowledge of knowledge. At the same time the definition of waste expands from an individual's perspective to a society's perspective. To become assessed at a society's level (metalevel), personal knowledge both at the base level and the metalevel needs to be shared by externalizing as *information* (see Figure.4-12(b), 4-13(b)).

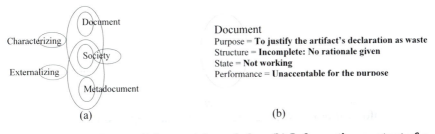

(a)   (b)

Figure 4-13. (a) Externalizing metaknowledge. (b) Information content of a metadocument.

Private ownership relation, once formed, is a restricted relationship. An owner of an artifact has no unlimited right to decide that the artifact has *no purpose in general*. This applies especially for individuals and small groups as private owners. Such an owner, doing so, would make decisions on behalf of potential other owners. An artifact could be declared terminally useless, and thus waste, on possibly selfish grounds to *get rid* of the artifact. How comprehensive, then, a human medium should be to have an unquestioned right as an owner to declare an artifact as waste? Should such decisions be made on local society's basis, nationally or globally? This must be dependent on the type and quantity of waste.

A private owner can freely argue that an artifact has *no purpose for him*, because he has no (more) use for it. What he then actually argues, is that there is no (more) justification for a given *event-medium composite* to survive For instance, although a newspaper/reader aggregate retains its potential to keep up *reading* activity going on even after everything has been thoroughly read, the rate of *information transfer* ultimately approaches zero. Beyond some point the composite does not perform satisfactorily in this respect any more. This state of affairs generates a need for the medium to do something about (manipulate) its own structure, if it aims to survive. In the newspaper/reader aggregate the active part is of course the reader. In principle his choices to manipulate are either to maintain the aggregate medium's topology and the associated ownership relation, or to cut the ownership relation to kill the composite as illustrated in Fig 10.

A newspaper is an artifact, which after being read is no more of use as an *information carrier* for its owner and is, when assessed from this perspective only, an obvious candidate for a waste. However, as discussed earlier, its information content remains intact on reading and about the same applies for the carrier material. A newspaper thus retains functionalities, which make its continued existence in the same or in another ownership relation justified. For instance, it can be stored or circulated for its information content. Suppose, that an owner decides to maintain the ownership relation for a while and to burn his newspapers in a fireplace. He uses his (assumed) right to manipulate the artifacts he owns. Upon burning a newspaper loses its identity. The medium, which originally had newspaper and reader as its primary constituents, splits and what was paper and ink (with embedded information) becomes an aggregate of materials including gaseous products and ash.

The materials formed on burning are still artifacts, but probably useless to the owner except possibly for their heat content. They are not artifacts in the sense of having purposively designed properties. Returned to the atmosphere and soil they become rehabilitated into natural products. At the same time the private ownership and the associated responsibilities dissolve, which appears to make burning a handy means for individuals of getting rid of old newspapers. In general the act of rehabilitating is not based on an individual's decision, but a decision of a society. This is because the decision should be based on shared knowledge. After the primary use of artifacts, not any exploitation of the remaining functionalities in a large scale is permitted by a society. In the case of used newspapers acceptable ways include de-inking and recycling to papermaking process, converting cellulose to other useful chemicals in various chemical or biochemical processes, and releasing the bond energy, stored in the molecular structure of lignocellulose, by incineration to make heat. Incineration gas and ash are unavoidable byproducts. If they cannot be disposed of as such, they are unwanted and thus waste. In individual cases, however, the fuzziness of the borderline between artifacts and natural objects may be exploited. For instance, after becoming a private owner of an amount of wood material, the owner has a right to manipulate (process) it. Upon processing, the material loses its original structure and identity, and some byproducts are generated, which cannot be utilized in the process. Because the byproducts have their origin in wood, there is a temptation to argue that they represent a part of the wood material that was actually never privately owned and thus remained natural and thus can be freely returned to where they came from. Untreated exhaust gas of combustion engines or discarded empty beer cans are two further examples of make-believe natural objects.

Reading a newspaper is not different from pouring milk form a container when viewed as an act of consuming the content of a packaging. The difference is in what is transferred: information in the former and material in the latter. At the end of consumption activity, a milk container is empty. It can be said to have less functionalities left than a newspaper, which still contains its information, and thus to have less justification for existence as an artifact. However, the information content would make a newspaper more valuable (less obvious candidate for waste) than a milk carton only, if it were a unique exemplar instead of just one of thousands of copies. If one copy is burned, the same information is still available in thousands of others. In this sense, newspaper can be regarded as a packaging and to belong to a different category than books for instance. As a packaging the possible constraints for destroying it are purely technical rather than ethical ones, which is opposite to what holds for books. We may conclude:

> An artifact having no purpose for its owner *can* be labeled as waste, if accepted by society.

The ultimate reason for an artifact to become a candidate for waste is its poor *performance*. Performance can appear poor when assessed against any of the criteria set and committed to by the designer, maker or user of an artifact. An artifact with poor performance *behaves* in a manner, which makes it unwanted to its owner. An artifact either lacks a required functionality or owns unwanted functionalities, which become manifest in unexpected situations. Poor performance of an artifact results from its *structure* being improper in the artifact's intended usage process. An improper structure can become manifest already before an artifact is subjected to use. An artifact may appear on inspection to be improperly designed or made. An initially proper structure can deteriorate in use due to wear or to being used improperly, and become, often unexpectedly, manifest as breakdown of the usage process. Obviously each of the processes of designing, making (manufacturing) and using artifacts can be responsible of poor performance. These processes are central for waste management and actually are the context, in which the concept of waste obtains its full meaning.

Including the potential for improving the processes of designing, making (manufacturing) and using of artifacts, the proposition for a definition of waste can be further narrowed. At the same time the definition of waste becomes inseparable from the definition of waste management. This is only natural by the same token as waste is inseparable from human. If there is no means for manipulating an artifact's purpose or structure to make its performance acceptable, we end up with the following definition:

> An artifact whose *performance* is *terminally* unacceptable *is* waste.

To express this definition in terms of conventional classes, let us return to the waste taxonomies proposed elsewhere. The traditional taxonomies classify wastes by *phase of material* (solid, liquid, gas), by *origin* (processing, household, packaging, or cleaning; construction and demolition; emissions treatment; energy conversion, etc.), or by *functionality* (inert, combustible, bio-degradable, hazardous, nuclear, etc.) [10].

Take a newspaper and a milk carton as examples. By traditional taxonomy, they are solid and combustible and originate from household and packaging. By the new definition, having passed through the consumption process they are artifacts left with no performance, when assessed by the owner against the *original* purpose. Being combustible they still may have an acceptable performance with respect to combustibility as a secondary functionality, if used as a fuel. If the owner does not want to use the artifact, having fulfilled the first purpose, for that secondary purpose, he obviously regards its performance unacceptable. But due to this secondary functionality (and possibly others), the performance is not terminally unacceptable. Newspaper and milk carton are not yet waste for the society. They are justified to survive at least in order to serve for burning as the secondary usage process, and for this (or some other) purpose their ownership should be traded. If the owner has a customer ready to accept the ownership and its associated responsibilities, the owner has a right to define the artifact as waste of his usage process (reading or milk consumption) and trade its ownership. What the original owner would get from this trade is the freedom from the ownership. The other alternative is of course that the owner recognizes some of the secondary functionalities himself and uses the artifact in a proper usage process, provided that it is accepted by the society.

The considerations above concerning the definition of waste can be summed up in a convenient format, following the waste taxonomy proposed [12]. Based on the *reason why* an artifact ends up becoming waste, the following four top-level classes arise:

Class 1: An unwanted unavoidable artifact created with *no purpose.*

Class 2: An artifact with *no secondary purpose* having fulfilled the primary purpose.

Class 3: An artifact with *unacceptable performance* assessed against its purpose.

Class 4: An artifact with acceptable performance, but *not used for the intended purpose.*

These classes represent the common denominators, which are argued to offer a holistic view into the field of waste management and a key for its systematization. When descending to the lower levels of taxonomy, the conventional classes of waste are encountered.

According to the taxonomy [loc.cit.] the following sub-classes of waste appear: 'To Class 1 belong outputs with negative market value, non-useful by-products, emissions, processing and process wastes, cleansing wastes, etc.; Class 2 is the group of single use or disposable products; most of the packaging, single use cameras, disposable diapers, etc.; Class 3 comprises non-functional, obsolete products, old furniture, discarded household appliances, non-rechargeable batteries, demolition waste, spoiled products, etc.; Class 4 contains products used in excess, or simply products that the owners do not wish to own anymore.'

By the PSSP approach the idea of what *can be* waste easily reaches beyond the traditional scope. In particular, consider human being itself as a process. An individual human being is a process, whose interior is an event-medium composite of mental/physico-chemical phenomena occurring in the mind/body medium. The products of this process are both of material and information type. Theoretically, when an individual is viewed in isolation, both the human process and its products are natural. When we view an individual, as we have to, in the context of the society, which is an artificial collective human process, also his outputs, while individually natural, turn to artifacts and thereby to candidates of waste. What is important here is to realize that also *information* is an artificial product and can have properties by which it may become labeled as waste. However, information waste and information waste management are not a part of the traditional waste management discourse.

It should be mentioned that *information waste* has not been any major issue under systematic scrutiny in the field of information science and technology either. Karabeg [8] proposes four criteria for assessing performance of information, one of which is nourishment. Using a food metaphor he argues that embedded in factual information there are 'information nutrients' and 'information poisons' that we usually are not aware of. Obviously 'poisonous' information would belong to Class 3 above. We leave the discussion of information waste outside the scope of this paper. Not because it were uninteresting, but because it deserves being discussed on a separate forum.

4.7 What is waste management?

4.7.1 Traditional view

Quoting earlier papers [10,11], in Article 1 the European Council Directive on waste defined waste management as: ''Waste management shall mean collection, transport, recovery and disposal of waste, including the supervision of such operations and after-care of disposal sites''. This definition of waste management has the same 'organizational' approach as the definition of waste. It is concerned with the existing amount of waste, trying to minimize the human-waste or environment-waste interface, to minimize potential impact. The Council Directive on waste prescribes that waste is to be recovered or disposed by means of the most appropriate methods and technologies to ensure a high level of protection for the environment and public health. In this context, 'the environment' means 'the whole of the natural world inhabited by living organisms, especially considered vulnerable to pollution'.

On the other hand, the 'waste management hierarchy', which is a widely accepted list of preferred waste management options, gives priority to waste minimization, which includes changing processes to prevent waste. Here, there is some conflict between the two interpretations of waste management. The definition of waste management suggests its role to be merely to get rid of existing waste, while the hierarchy suggests that, ideally, we should avoid having (producing) waste. How then shall we understand waste management?

Semantically, the expression is an interesting use of words. 'To manage' is, according to the Merriam Webster On-line Dictionary ''to handle or direct with a degree of skill, to work upon or try to alter for a purpose, to succeed in accomplishing or to direct or carry on business or affairs''. While 'management' is defined as: ''judicious use of means to achieve an end''. It appears from these definitions, and it is also our understanding of management as presented in an earlier paper [13] that management is control of activities, while the expression of 'waste management' syntactically suggests that it is control of materials.

4.7.2 Waste management in PSSP format

Possibility for a causal relationship to be established between an event and a medium is the pre-requisite for an event-medium composite to build up. Ownership is a right or responsibility of an event-medium composite to monitor and manipulate itself. Upon manipulation, properties of both the medium and the event change observing some causal law. When the medium is homogeneous in the sense that it is not an aggregate of (spatially) separate sub-media, the right (at least in principle) is unlimited like when homogeneous material approaches thermodynamic equilibrium (entropy principle of Nature) or an individual's (embodied) mind reorganizes itself in thinking (principle of freedom of thought). In the domain of waste management the medium is a heterogeneous aggregate of human and artifact. Because human being cannot be defined in isolation, but always as a member of a society, the ownership over an artifact and the associated right and responsibility to manipulate it is under social control. This makes waste management ultimately a social rather than an individual activity.

The PSSP-based classification of wastes is a sound starting point for systematizing waste management. Taxonomy on the basis of why an artifact ends up becoming waste will yield classification of activities aiming either to eliminate reasons for artifacts to turn to waste of any of the four types, or to do something about already existing wastes of these four types.

Waste under Class 1 is created as an unwanted but not avoided output with no purpose. Most technical processes aim at specific products and can seldom avoid producing unintended, often undesired, by-products that we call waste. This applies to all industrial processes and technical devices, especially those using fossil fuels. In these processes the material manipulated typically undergoes a profound structural change. The products and by-products may be either material or energy or both. For instance, automobile exhaust gas is an unavoidable purposeless material by-product of a process (or device) producing mechanical energy (via heat energy) as its product (see Figure.4-14).

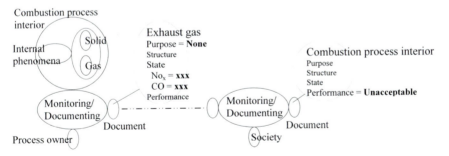

Figure 4-14. Society's stand on exhaust gas emissions.

At the highest level of waste management the *unavoidability* of an unwanted output should be questioned. Can the process or device be *designed, built, retrofitted* and *operated* such that an unwanted output is eliminated? If the answer is no, can the output be reduced? If not to the extent, which satisfies the society, then the process or device may not have justification for existence (Figure.4-15). However, an unwanted output of one process may be a valuable input for another. A viable solution might then be found by searching for possibilities of *waste-trade*.

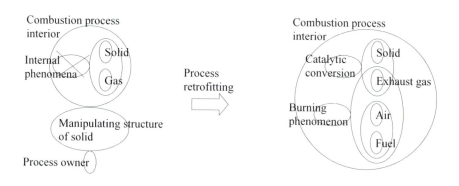

Figure 4-15. Waste management by retrofitting to reduce harmful components' release.

Waste under Class 2 forms after an artifact has fulfilled its *single* intended purpose. The best example of this category is packaging. It is already at the design phase that waste management should take the responsibility of the fate of packaging after it has fulfilled its primary purpose. The responsibility is taken by designing packaging to have appropriate secondary functionalities embedded in its structure. When the most probable fate is to end up in *landfill*, proper secondary functionalities would include such as low weight, small volume, collapsible shape and flexible walls. If material *recovery* seems feasible, it is essential to use materials that are the most economical to recycle. If *incineration* is planned, it is vital to omit chemicals that may lead to toxic emissions. Incineration as an option of waste management is illustrated in the Case to follow.

Waste under Class 3 is created, when an artifact does not *behave*, as it is required. As discussed earlier, each of the processes of designing, making (manufacturing) and using artifacts can be responsible of poor performance. From an artifact's life cycle perspective waste management should focus on an artifact's *design* stage and seek to produce goods with a maximum lifetime. Most products, however, have a discrete lifetime and after this time, often cease to be useful. Design should focus on the ease of *assembly* and *disassembly* so that even if the product as a whole ceases to be useful, some parts of it still can be utilized, as illustrated in Figure.4-16.

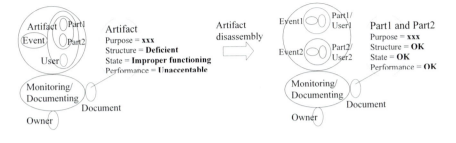

Figure 4-16. Waste management by ideal design: Useless artifact disassembles into useful parts.

Waste under Class 4 is created because its owner does not use an artifact for its intended purpose. The discussion in [12] illustrates the kinds of waste management activities, which arise in this category. 'Gourlay considers the small amount of mustard left on a plate at the end of a meal. This is neither useless, nor has it lost its properties. It has become waste because the owner failed to consume it. Gourlay also points out other cases, such as agricultural production and fish farming, when substances that fail to reach their target (nitrates leaching into soil, food and chemicals fed to fish ending up at the bottom of the sea) are wasted not because the owner failed to use them, but because he used them in excess (see Figure.4-17). Both of these waste types defy 'the owner discards or wants to discard' class of definition. Did the farmer want to discard useful fertilizer? Did the fish-farmer have any intention of discarding perfectly useful fish food? Clearly not, but the fish food dispersed to the bottom of the sea is unavailable to the fish; the fertilizer washed down the watercourses is wasted and non-retrievable by any means. Why did they become waste?'

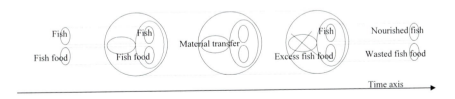

Figure 4-17. Example of waste formation due to excessive use of artifact.

Waste of the above type invites waste management to focus on *legal*, *educational* and *ethical* issues. While there is a possibility of controlling consumers through legislation or by the use of motivational tools, the best way to influencing people is by raising consumer awareness. By increasing their knowledge through education, consumers become aware of their actions and the possibilities and responsibilities in environmental protection. Legislation is essential, but the greatest gains will be achieved through a well-informed, environmentally conscious, ethical public.

4.8 A case

Let us start by considering the natural ownership relation between paper and the surrounding air. There is a potential for some physico-chemical phenomena to occur in this medium but usually the preconditions are minimal and practically nothing happens. When used newspapers keep on accumulating, the medium's inertness creates a problem, which soon becomes a waste management issue. Society needs to make a decision what to do with accumulating newspapers. Various choices exist. We do not consider in this paper whether some alternative might better than another. From the point of view of demonstrating the power of the PSSP formalism, any choice will do. So, we pick up *incineration*.

The PSSP formalism does not pose any restriction for where to start modeling the case. So lets start from an easy end that has been treated already earlier, a person reading a newspaper. As discussed, the person and the newspaper make up an aggregate medium in which the reading activity advances. After some while the reading person feels that he has extracted everything essential from the information content and ceases the activity (see Figure.4-18). From that on the paper is useless for him. It has become a candidate for waste. The problem to solve is, from all such readers' perspective, how to get rid of the old newspapers.

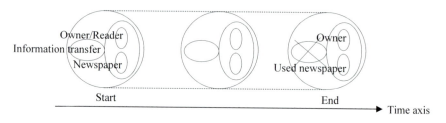

Figure 4-18. Progress of information transfer from newspaper to reader on reading

While in theory all used newspapers could be burnt in private fireplaces by individual owners, it cannot be the ultimate newspapers incineration method. Incineration needs to be done under society's control. One important reason for that is the risk of pollution due to harmful releases. That is why, in modern society, strictly controlled incineration plants are the preferred solution. This would give an opportunity for individual newspaper owners to trade the ownership with an incineration company (Figure 4-19).

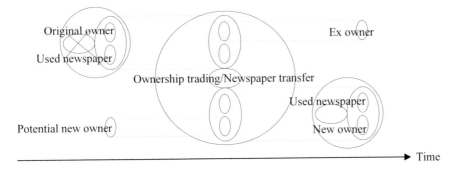

Figure 4-19. Trading ownership and transferring used newspapers to a new owner.

Taking a view from the other end of the case, a human (or social) medium, which is an aggregate of the incineration process owner and the rest of the society, keeps up an activity, which can be called *social control*. It legalizes the ownership transfer concerning used newspapers. It poses conditions for, and keeps the society informed about, the incineration business and its social and environmental impacts. The process owner, meaning the whole personnel responsible of running the business, acts as a link between the society and the incineration process (Figure.4-20).

Figure 4-20. Linking social control and process control.

Prior to landing to incineration as the waste management option, the society has *identified* the used newspapers as waste of Class 2, that is, artifact having fulfilled its primary purpose and not having been assigned a secondary one. The decision to make at this stage is whether to declare the material as a natural product, which of course is not a realistic choice, or to assign it a secondary purpose based on the remaining functionalities. The decision should be based on *knowledge*. Having knowledge is being able to build scenarios for, or simulate, the various options. In PSSP terms, knowledge is being able to *specify* the attributes of each of the objects involved. What is especially needed, is knowledge of the causalities/rationales of all the relevant events involved, both physico-chemical phenomena and human activities.

Although incineration is a complex *process*, we may still approach its description by starting from the simple event-medium composite depicted in Fig.19, to the right. However, a much more detailed description of the composite, than in the case of reading a newspaper, is necessary. There are important sub-events, whose description implies including in the model several other sub-media. The medium for the burning phenomenon is the medium of the incineration process *interior* and includes, besides used newspaper, also the other participating material components, both the reactants and the residues. The medium of the incineration process *exterior* is that of the whole 'infrastructure', which directly or indirectly *interacts* with the incineration process interior. Direct interaction includes material transfer in the form of input of newspaper and air, and output of gaseous products and ash. The external sub-media in direct interaction include newspaper (as feed), atmosphere, soil, water (as heat receiving material) and people. Indirect interaction includes the control activities shown in Fig.20. Upon detailing the event-medium composite becomes as depicted in Figure.4-21.

Detailing by disaggregating, as done in Figure.4-21, does not yet make explicit the *roles* of sub-media either as the medium of process interior or process exterior. This is, however, what we want to do when taking a holistic process view into incineration as a waste management option. Regrouping sub-events and sub-media as belonging either to incineration process interior or exterior, a more illustrating formal description follows (Figure.4-22).

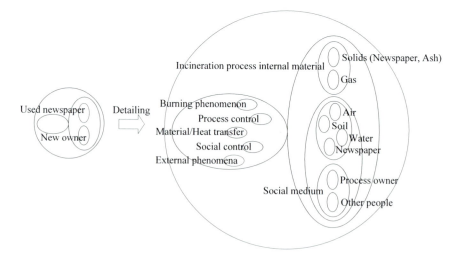

Figure 4-21. Incineration (a waste management option) described as a single event-medium composite.

In the PSSP model depicted in Fig.22, the members of the society, that is, people around the incineration plant, are in two roles: as an external medium and as a controlling medium. In the former role people are among the external material and take part in the interaction between the incineration process interior and exterior. They are agents mediating the newspaper feed of the plant and also agents who experience the effects of the plant's material and energy outputs. The same people in the role of a controlling medium are the ones, who are ultimately responsible of having made the choice to implement this waste management option. It is thus the society as a whole, which is responsible of accepting that the material output of the incineration plant can be taken as natural products and as such natural and harmless components of atmosphere, soil and people. 'Harmless' means that no unwanted spontaneous phenomena would occur in the external medium with the result that people can breath air and cultivate soil as before without fear of risking their health or without need to stand unpleasant odors, noise or other nuisance.

Society can be viewed as a pseudo-homogeneous medium having a right and responsibility to monitor and manipulate itself. A society's ownership over itself is not unlimited in the same sense as it is for truly homogeneous media like homogeneous matter or individual human mind. In manipulating itself a society needs to take the interests of individual people or of groups of people into account, that is, to observe the principle of democracy. Society as a pseudo-homogeneous human medium is similar to individual human being in its capability to act at more than one level at the same time: both as a base level actor and as a controlling actor. Both an individual and a society in the self-controlling role pose goals to themselves and monitor, and allocate resources for, their own progress as base-level actors in pursuing the goals.

In Figure.4-22, the process owner, despite being depicted as a separate medium, is a base-level actor of the society, when the latter is viewed as a pseudo-homogeneous medium. Thus the process owner is also among the decision-makers when it comes to the choice and the conditions of incineration as a waste management option for used newspapers. The process owner, again despite being depicted as a separate medium, acts at the two levels also as a part of the incineration process exterior: as a base-level actor, and simultaneously as a decision maker when it comes to operating conditions of incineration, both as a technical process and a business process.

We may conclude that the private ownership over used newspapers migrates from individuals to incineration process owner and ultimately to the society. It is the society, which blesses the waste management decisions and declares the outputs of the incineration plant as natural products. If a society makes a fatal mistake here, no one can help, if it is not Nature.

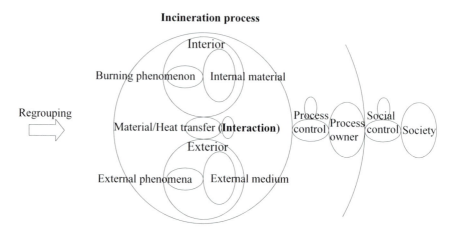

Figure 4-22. Incineration process as a waste management option.

4.9 Discussion

In PSSP ontology waste management is a process. It is a process, because anything in reality having a boundary can be taken as a process. Waste management has a boundary, because it is restricted to concern events and media related to human civilization. The realm of waste management has various extensions. In one case the effects of waste related activities are local, in another case far-reaching. It is up to the society to set the boundaries in each case. This is done based on the knowledge, ignorance and attitudes a decision-maker has. The worst decisions are those based on blindness or ignorance concerning what a decision-maker should know for a decision to be responsible. The only remedy for that is to have a holistic view based on a holistic representation of waste management as a process including all relevant technical, social and ethical issues in an integral format. Such a representation is possible only when built upon an adequate ontology. No domain ontology would do for two reasons. First, there is no predefined domain of waste management in the sense that the issues, which can be relevant for waste management, cannot be exhaustively listed. Second, the issues, which are known to be relevant for waste management, reside in various conventional domains and are loaded with conventional domain specific meanings, which unfortunately is often an unsurpassable obstacle for knowledge integration.

PSSP ontology has been offered as an approach to ignorance management [14]. As such it is also a tool for formalizing, and an approach to systematizing, waste management.

4.10 Acknowledgement

The comments of Dr. Eva Pongrácz are gratefully acknowledged.

4.11 References

1. Lowe, E.J., Recent Advances in Metaphysics, *Facta Philosophica* **5** (2003) 3-24.
2. Pohjola, V.J., *POEM Guide Book. Introduction to SHE conscious process design.* University of Oulu Press, 2001, Oulu, Finland.
3. Pohjola, V.J., Fundamentals of safety conscious process design. *Safety Science* **41** 2-3 (2003) 181-218.
4. Pohjola, V.J., Ontology supporting knowledge integration, in Milutinovic, V. and Vujovic, I. (eds) *Advances in the Internet Technology: Concepts and Systems*, IPSI and Academic Press, Belgrade, 2004, p. 146-159.
5. Korzybski, A., *Science and Sanity*, The International Non-Aristotelian Library Publishing Company, 1980.
6. Bohm, D., *Wholeness and the implicate order*, Routledge and Kegan Paul, 1980, London.
7. Pongrácz, E., Pohjola, V.J., The importance of the concept of ownership in waste management. Proceedings of the 15th International Conference on Solid Waste Technology and Management, December 12-15, 1999, Philadelphia, PA.
8. Karabeg, D., Information design – an informing for the 21st century. Proceedings of IPSI-2004 Stockholm, September 24-26, 2004, Stockholm, Sweden.
9. Smith, B., Varzi, A.C., Fiat and bona fide boundaries, *in* SC Hirtle. S.C. and Frank, A.U. (eds), Spatial Information Theory, COSIT '97, 1997, Laurel Highlands, PA.
10. Pongrácz, E., Pohjola, V.J., Re-defining waste, the concept of ownership and the role of waste management. *Conservation and Recycling* **40** (2004) 141-153.
11. Pongrácz, E., Re-defining the concepts of waste and waste management: Evolving the Theory of Waste Management. Doctoral dissertation. University of Oulu, 2002, Finland.
12. Pongrácz, E., Pohjola, V.J., The conceptual model of waste management. Proceedings of the ENTREE'97, November 12-14, 1997, Sophia Antipolis, France, p. 65-77.
13. Pongrácz, E., Pohjola, V.J., Object-oriented modeling of waste management. Proceedings of the 14th International Conference on Solid Waste Technology and Management, November 1-4, 1998, Philadelphia, PA.

14. Pohjola, V.J., Ignorance management by contextual templates. Proceedings of IPSI-2004 Stockholm, September 24-26, 2004, Stockholm, Sweden.

5 Formal Methods for Modeling Biological Regulatory Networks

Author
Adrien Richard, Jean-Paul Comet and Gilles Bernot
CNRS & Université d'Évry, France

Summary
This chapter presents how the formal methods can be used to analyse biological regulatory networks, which are at the core of all biological phenomena as, for example, cell differentiation or temperature control. The dynamics of such a system, i.e. its semantics, is often described by an ordinary differential equation system, but has also been abstracted into a discrete formalism due to R. Thomas. This second description is well adapted to state-of-the-art measurement techniques in biology, which often provide qualitative and coarse-grained descriptions of biological regulatory networks. This formalism permits us to design a formal framework for analysing the dynamics of biological systems. The verification tools, as model checking, can then be used not only to verify if the modelling is coherent with known biological properties, but also to help biologists in the modelling process. Actually, for a given biological regulatory network, a large class of semantics can be automatically built and model checking allows the selection of the semantics, which are coherent with the biological requirement, i.e. the temporal specification. This modelling process is illustrated with the well studied genetic regulatory network controlling immunity in bacteriophage lambda.

5.1 Introduction

Biological systems are one of the most fascinating aspects in biology. They control such diverse dynamics phenomena as temperature control in warm-blooded animals; differentiation of a zygote into the various specialized organs, tissues and cells of the mature organism; the fate of certain viruses, called temperate bacteriophages, which upon infection of a bacterial population can behave in two extremely different ways. Most of these infected cells display a response called *lytic*: virus multiplies and kills cells. But, a fraction of the cells become *lysogenic bacteria* and carry the viral genome in a dormant form making the host immune towards infection of other virus.

Hossam A. Gabbar (ed.), Modern Formal Methods and Applications, 83–122.
© 2006 *Springer. Printed in the Netherlands.*

Computational systems biology tries to establish methods and techniques that enable us to understand such systems as systems, including their robustness, design and manipulation. It means to understand : the structures and the dynamics of systems, methods to control, design and modify systems to cope with desired properties. The modelling contributes in a major way to reach these aims by introducing methods for understanding, simulating and predicting the behaviour of the systems. However, the modelling of biological systems is currently subject to two major difficulties: the biochemical reaction mechanisms underlying the interactions of systems are usually not or incompletely known and quantitative information on kinetic parameters or molecular concentration is rarely available. Thus the modelling activity needs an interaction with the experimental biology to confront models to biological objects. Consequently as in the design of large computing systems, two activities can be distinguished in the modelling step:

1. Build a rigorous model of the system satisfying the assumed behaviour corresponding to biological knowledge,
2. Design experiments to verify *a posteriori* the model predictions.

Here we would like to show that some methods from computer science can be reused in the context of system biology, as, for example, formal methods for validation and verification used for the design of large computing systems. For designing experiments, we just mention that the test methods *via* test generation from model theories may be an efficient way to propose experiments permitting biologists to validate or refute models. For building a rigorous model, the model checking verification tool is particularly suited. In this chapter we present an application of this formal method to build qualitative models of biological regulatory networks.

A biological regulatory network describes interactions between the biological entities, often macromolecules or genes, of a given system. It is statically represented by an interaction graph whose vertices abstract biological entities and arcs their interactions. For describing the evolution of the system, the concentration level of each entity is represented by a value associated to the corresponding vertex. The temporal evolution of these levels constitutes the dynamics of the system.

Ordinary differential equation systems have been first used for describing the dynamics of networks. They are powerful tools particularly to model metabolic processes [33]. However, due to the non-linearity of biological regulations, these differential equation systems cannot often be solved analytically. They can be solved numerically to any desired precision, but this precision itself may be misleading because the values of the parameters and the shape of interactions often have to be guessed for lack of experimental data. This remark led Thomas to simplify the models: he introduced in the 70's a

Boolean approach to capture the qualitative nature of the dynamics and he proved its usefulness in the context of immunity in bacteriophages [30, 27]. Later on, he generalized his formalism to multi-valued levels of concentration (the so called multi-valued logic or ``generalized logical approach'' [32]) since the Boolean idealization may be too caricatured to correctly model biological systems. It has been proved that this qualitative description allows the representation of the essential qualitative features of an ordinary differential equation system provided that the differential equations are piece-wise linear [22]. The underlying parameters of the qualitative description can be deduced from the kinetic parameters of the continuous system but can take only a finite number of values. Consequently, all possible qualitative features of the system can be reduced to a finite number of models *i.e.* parameterisations.

Certainly the most important concepts of the generalized logical analysis are those of positive and negative feedback circuits. If an entity tends to favour (resp. decrease) its own production *via* the feedback circuit, the circuit is said positive (resp. negative). It has been conjectured by Thomas [28] and then proved in different contexts [18, 23, 5, 6, 25] that at least one positive circuit is necessary to generate multi-stationarity whereas at least one negative circuit is necessary to obtain a stable oscillatory behaviour. These concepts are especially important since when modelling biological systems, differentiation and homeostasis have often to be taken into consideration. In such cases, these biological constraints can reduce drastically the set of models to consider. These properties can be reinforced by introducing some more complex properties on the dynamics of the system extracted from the biological knowledge or hypotheses. It becomes necessary to construct models which are coherent not only with the previous conditions of multi-stationarity/homeostatis but also with the additional ones. Formal methods from computer science should be able to help modeller to automatically perform this verification [3, 17] and to select exhaustively all suitable models.

The chapter is organized as follows. Section 5.2 introduces the formalism due to Thomas for modelling the dynamics of a biological regulatory network. The resulting dynamics corresponds to a Kripke structure, which can be deduced easily from the interaction graph. Section 5.3 describes the link between this transition system and the dynamics obtained with the classical modelling using piece-wise linear ordinary differential equation systems. Section 4 explains how formal methods can improve the modelling process of regulatory networks. The temporal properties have first to be translated into a temporal specification language. Then one has to answer automatically the question: does a given model satisfy the given temporal specifications? Model checking makes this stage automatic and its principle is also presented in this section. Section 5 illustrates the use of model checking to model the well studied genetic regulatory network of temperate bacteriophage lambda rapidly described before. A model of this system has already been proposed by Thieffry and Thomas in [26]. We show that our approach, using model checking, automatically selects this model as well as other models satisfying the same criteria of validation.

5.2 Qualitative dynamics of biological regulatory networks

The multi-valued modelling of Thomas is able to represent the qualitative dynamics of biological regulatory networks whose entities can be molecules, macromolecules, cells, organs, or organisms, if no societies. In fact, all systems whose regulations have a sigmoid shape can be modelled in this formalism. The regulations of genetic regulatory networks have almost always a sigmoidal nature that explains why this formalism has been introduced in this context and why its main application domain remains the genetic regulatory networks. In such systems, the concentration of a protein encoded by a gene u may activate or inhibit the synthesis of proteins encoded by other genes or itself (figure 5-1).

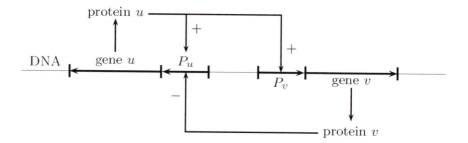

Figure 5-1. A genetic regulatory network. The gene u synthesizes a protein which activates the expression of gene v and itself by binding the promoters P_v and P_u respectively. In turn, the protein of gene v inhibits the expression of gene u when it binds P_u. Then, the arrow from a gene to its protein represents the transcription and translation processes and the arrow from a protein to a promoter abstracts the diffusion and the fixation of the protein on the promoter.

If the protein of u activates (resp. inhibits) the expression of a gene v, we said that u is a positive (resp. negative) regulator of v. In such situation an increasing of the concentration of the protein encoded by u induces an increasing of the rate of synthesis of the protein encoded by v. Generally, the relation between the concentration of a regulator and the rate of synthesis of its target is, as we have seen before, sigmoidal. When the sigmoid is steep, as in figure 5-2-(a), u has a little effect on v if it is below the concentration threshold θ_{uv} and at higher concentration a plateau is reached representing the maximal rate of synthesis of v under the effect of u. Naturally, for an negative regulator, the sigmoid is decreasing.

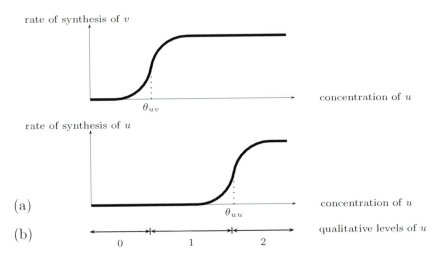

Figure 5-2. (a) Sigmoid relations between the concentration of *u* and the rate of synthesis of *v* and itself. As *u* is an activator of *v* and itself, see figure 5-1, both sigmoids are increasing. (b) Resulting qualitative levels of *u*.

This section presents successively how the regulations can be summarized into a regulatory graph corresponding to the static part of the modelling, then introduces the parameters, which encode the effects of regulators on their targets in all possible situations, and finally presents how the dynamics can be deduced from these parameters.

5.2.1 Biological regulatory graphs

To formally define the static part of biological regulatory networks, we use labelled directed graphs. Vertices represent the biological entities of the network and arcs their regulations.

Definition 1 [Biological regulatory graph] *A biological regulatory graph is a labelled directed graph $G=(V,E)$ where*

- *each vertex v of V, called* variable, *is provided with a boundary $b_v \in$ IN less or equal to the out-degree of v in G.*
- *each arc $u \to v$ of E is labelled with a couple (t_{uv}, α_{uv}) where t_{uv} is an integer between 1 and b_v, called* qualitative threshold *and where $\alpha_{uv} \in \{+,-\}$ is the sign of the regulation.*

Moreover it is required that for any variable u with $b_u > 0$, $\forall\ i \in \{1,2,...,b_u\}$, there exists a successor v of u such that $t_{uv} = i$.

In a biological regulatory graph G, the set of the regulators of a variable v corresponds to the set of its predecessors in G, denoted by $G^-(v)$, and the set of its targets corresponds to the set of its successors in G, denoted by $G^+(v)$. Each regulation $u \rightarrow v$ is labelled by a sign α_{uv}, which indicates if u is an activator or an inhibitor of v, and by a qualitative threshold t_{uv}. Thresholds t are integers and do not correspond to biological thresholds $\theta \in \mathbb{R}$, most often difficult to measure, but they give the order of the continuous thresholds: if $t_{uv} = i$ then the corresponding continuous threshold θ_{uv} is the ith lowest threshold among $\{\theta_{uv} \mid v \in G^+(u)\}$. That explains the requirement on qualitative thresholds of the previous definition, which implies that b_v is the number of different thresholds "outgoing" from v.

Figure 5-3-(a) gives an example of biological regulatory graph, which can be deduced from the genetic regulatory networks described in figure 5-1. Figure 5-2 assumes that $\theta_{uv} < \theta_{uu}$, and consequently $t_{uv} = 1$ and $t_{uu} = 2$.

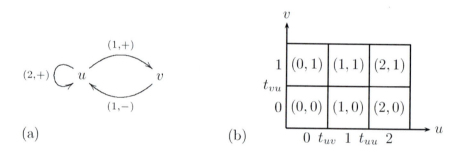

(a)   (b)

Figure 5-3. (a) Biological regulatory graph deduced from the genetic regulatory network of figure 5-1. (b) States of the biological regulatory graph.

Obviously concentration levels are associated to variables. For describing the evolution of the concentration level of each variable, it is necessary to know which regulators have an effect on the variable. Only the position of the regulator concentration with regard to their thresholds is sufficient. The concentrations are then discretized according to thresholds and each variable can take a finite number of values called abstract qualitative levels. For example, in figure 5-2-(b), the variable u has three different behaviours with regard to its targets:

- In the first region (the concentration of u is less than θ_{uv}), u acts neither on v nor on itself.
- In the second region (the concentration of u is between θ_{uv} and θ_{uu}), u acts on v but it still not act on itself.

- In the last region (the concentration of u is greater than θ_{uu}), u acts both on v and on itself.

Three qualitative levels emerge, 0, 1 and 2, corresponding to the three previous regions and constitute the only relevant information from a qualitative point of view. More generally, a variable v can take b_v+1 qualitative levels, from 0 to b_v, and the qualitative level q means that v acts on all targets v' such that $t_{vv'} \leq q$. A state of the system is then defined as a vector constituted by qualitative levels of variables.

Definition 2 [Qualitative state] *Let $G=(V,E)$ be a biological regulatory graph. A qualitative state of G is a vector $q=(q_v)_{v\in V}$ such that for all $v\in V$, $q_v \in \{0,1,...,b_v\}$. The set Q of states of G is then defined by $Q=\prod_{v\in V}\{0,1,...,b_v\}$.*

In the sequel, we write $v = l$ for denoting $q_v = l$ if it does not cause confusion. Figure 5-3-(b) shows the possible states of the biological regulatory graph of figure 5-3-(a).

5.2.2 Models of biological regulatory graphs

Remember that the sigmoid nature of a regulation $u \to v$ leads to distinguish two different situations: if $u \geq t_{uv}$, then the regulation is active and if $u < t_{uv}$, it is not. We said that u is a *resource* of v when u induces an increasing of v:

- If the regulation $u\to v$ is positive, u is a resource of v when the regulation is active,
- If the regulation $u\to v$ is negative, u is a resource of v when the regulation is not active.

From the point of view of resources, the absence of an inhibitor acts as the presence of an activator.

Definition 3 [Resources] *Let $G=(V,E)$ be a biological regulatory graph, $v\in V$ and $q\in Q$. The set $\omega_v(q)$ of resources of v at the state q is the subset of $G^-(v)$ defined by:*
$$\omega_v(q)=\{u\in G^-(v) \mid (q_u \geq t_{uv} \text{ and } \alpha_{uv} = +) \text{ or } (q_u < t_{uv} \text{ and } \alpha_{uv} = -)\}.$$

At the state q, the evolution of variable v depends on its resources $\omega_v(q)$. It remains to define in which direction evolves v at the state q. The parameter $K_{v,\omega_v(q)}$, called the attractor of v when the resources are $\omega_v(q)$, denotes the level towards which v is attracted:

- if $q_v < K_{v,\omega_v(q)}$ then v tends to increase,

- if $q_v = K_{v,\omega_v(q)}$ then v does not evolve and

- if $q_v > K_{v,\omega_v(q)}$ then of v tends to decrease.

Different values for these parameters are possible and we call *model* of a biological regulatory graph, a possible instantiation of these parameters.

Definition 4 [Model of a biological regulatory graph]   *Let $G=(V,E)$ be a regulatory graph. A model of G, denoted M(G) by abuse of notation, is a family of natural numbers $K_{v,\omega}$ indexed by the set of couples (v,ω) such that*

- $v \in V$,

- $\omega \subseteq G^-(v)$,

- $K_{v,\omega} \leq b_v$.

It is often additionally required that:

$$K_{v,\omega} \leq K_{v,\omega'} \quad \text{for all} \quad v \in V, \quad \text{and} \quad \text{for all} \quad \omega,\omega' \subseteq G^-(v) \text{ such that}$$
$\omega \subseteq \omega'$.   (1)

These constraints, called Snoussi's constraints in the remainder, mean that the more a variable has resources the greater is the level towards which it is attracted. In other words, neither the presence of an activator nor the absence of an inhibitor can induce a decrease of the considered target (see the following section for the mathematical grounds of these constraints). This property, as well as signs of regulations, can often be deduced from biological knowledge and when it can be used, the number of models to consider for a given biological regulatory graph decreases drastically.

Table 5-1. One possible model for the biological regulatory graph of figure 5-3. The table gives for each state the corresponding attractors and tendencies deduced from the model.

Model		u	v	Attractors				Tendencies	
$K_{u,\{\}}$	$=0$	0	0	$K_{u,\{v\}}$	$=2$	$K_{v,\{\}}$	$=0$	↗	→
$K_{u,\{u\}}$	$=2$	0	1	$K_{u,\{\}}$	$=0$	$K_{v,\{\}}$	$=0$	→	↘
$K_{u,\{v\}}$	$=0$	1	0	$K_{u,\{v\}}$	$=2$	$K_{v,\{u\}}$	$=1$	↗	↗
$K_{u,\{u,v\}}$	$=2$	1	1	$K_{u,\{\}}$	$=0$	$K_{v,\{u\}}$	$=1$	↗	→
$K_{v,\{\}}$	$=0$	2	0	$K_{u,\{u,v\}}$	$=2$	$K_{v,\{u\}}$	$=1$	→	↗
$K_{v,\{u\}}$	$=1$	2	1	$K_{u,\{u\}}$	$=2$	$K_{v,\{u\}}$	$=1$	→	→

Models $M(G)$ of the biological regulatory graph G of figure 5-3 are all possible instantiations of six parameters: $K_{u,\{\}}$, $K_{u,\{v\}}$, $K_{v,\{\}}$, $K_{v,\{u\}}$, $K_{v,\{v\}}$, $K_{v,\{u,v\}}$. Because $b_u = 1$ (resp. $b_v = 2$) each $K_{u,...}$ (resp. $K_{v,...}$) can take 2 (resp. 3) different values. So $2^2 \times 3^4 = 324$ different models can be *a priori* associated to G, but only 60 of them respect the Snoussi's constraints. Table 5-1 gives the tendencies of variables resulting from such a model.

More generally there are $\prod_{v \in V} (b_v + 1)^{2^{|G-(v)|}}$ models associated to a biological regulatory graph G. This number increases exponentially with the number of predecessors of each variable and even if static constraints on parameters are used, as the Snoussi's constraints, it remains huge. Moreover, since parameters K are most often not measurable *in vivo*, additional properties deduced from biological experiments are needed to eliminate the models whose dynamics do not satisfy them.

5.2.3 Dynamics of models

The classical approach to describe the dynamics of models is to define the state of the system at time $t+1$ from its state at time t. One possibility is to consider that the next state is directly the attractor of the current state: if q is the current state then $q' = (K_{v,\omega_v(q)})_{v \in V}$ is the next one and we said that there is a transition from q to q' (figure 5-5-(a)). This description raises serious problems for its application to biological systems:

1. From any initial state, the system will follow a well-defined path, without any branching or possibility of choice whereas biological systems typically include choices among several pathways (as illustrated for example by the numerous different pathways leading to various cell lines from a zygote during embryonic development).

2. Suppose that v is a gene which can take two values ($b_v = 1$) and that the current state is q. If $q_v = 0$, then $K_{v,\omega_v(q)} = 1$ means that resources of v induce the production of the corresponding protein. This protein will appear after a time delay corresponding, for example, to the time of diffusion of its regulators (figure 5-4). Similarly the same phenomenon is observed when $q_v = 1$ and $K_{v,\omega_v(q)} = 0$ with an *a priori* different delay. However, when q differs from $q' = (K_{v,\omega_v(q)})_{v \in V}$ by at least two components, the corresponding variables change simultaneously (dashed arrow in figure 5-5-(a)). This synchronous description thus assumes that time delays are equal which is unlikely.

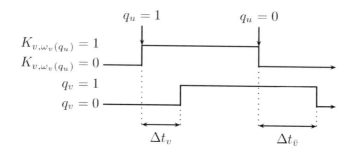

Figure 5-4. Time delays. Gene *v* has a unique regulator *u* which is an activator. Initially, both *u* and *v* are absent. Then, protein *u* appears and stimulates the expression of *v* ($K_{v,\omega_v(q_u)} = K_{v,\{u\}} = 1$). The resulting protein appears after the delay Δt_v. Finally, the protein *u* disappears, the gene *v* is no more stimulated ($K_{v,\omega_v(q_u)} = K_{v,\{\}} = 0$), and the protein *v* disappears after the different delay $\Delta t_{\bar{v}}$.

3. If the attractor of a variable is sufficiently away from its current value, one can have $|q_v - K_{v,\omega_v(q)}| > 1$. In such cases the qualitative level increases abruptly and jumps several thresholds (dotted arrow in figure 5-5-(a)). Since the dynamics of the model abstracts a continuous phenomenon, during a transition, each variable can pass through at most one threshold.

These points lead us to introduce the following asynchronous description.

Definition 5 [Asynchronous state graph]   *Let $G=(V,E)$ be a biological regulatory graph and M(G) be a model of G. The asynchronous state graph of M(G) is a directed graph whose set of vertices is the set Q of states of G, and such that there is an edge from q to q' if:*

*   *For all variables $v \in V$, $q_v = q'_v = K_{v,\omega_v(q)}$ or*
*   *There exist a variable $v \in V$ such that:*
    *   *For any variable $u \neq v$, $q_u = q'_u$ and*
    *   *$q_v < K_{v,\omega_v(q)}$ and $q'_v = q_v + 1$ or $q_v > K_{v,\omega_v(q)}$ and $q'_v = q_v - 1$.*

In this definition, a state *q* which has itself as successor, is a *stable steady state* of the asynchronous state graph: $q_v = K_{v,\omega_v(q)}$ for all $v \in V$. Otherwise, if *q* is a state for which *n* variables tend to evolve (*n* variables *v* such that $q_v \neq K_{v,\omega_v(q)}$), *q* has *n* successors and each of them differs from *q* by only one component corresponding to one of these *n* variables. Thus, when time delays are unknown, the asynchronous state graph contains all the *a priori* possible transitions. Some of them can be removed when time delays are taken into consideration.

Figure 5-5 shows the synchronous and asynchronous dynamics of the model of table5-1. The attractors are the same in both descriptions but paths differ: the asynchronous state graph contains a circuit $(0,0) \rightarrow (1,0) \rightarrow (1,1) \rightarrow (0,1) \rightarrow (0,0)$, which is absent in the synchronous description.

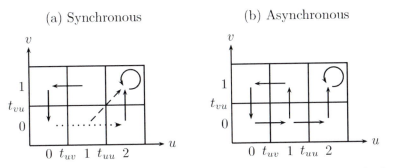

Figure 5-5. Synchronous and asynchronous dynamics for the model, given in table 5-1, of the biological regulatory graph of figure 5-3.

5.3 Differential modelling

We have seen that the asynchronous dynamics is more suited than the synchronous one for describing biological regulatory networks. This section proves how this asynchronous description can be deduced from a discretization of a particular class of ordinary differential equation systems classically used for describing biological regulatory networks.

5.3.1 Ordinary differential equation systems

Classically, the dynamic of a biological regulatory graph $G=(V,E)$ is modelled by ordinary differential equation systems [10, 22, 29], called here *underlying differential systems* (UDSs for short) of G, whose form is

$$\dot{x}_v = f_v(x) - \lambda_v x_v, \quad \text{for all } v \in V,$$

where $x = (x_v)_{v \in V}$ is a vector, whose components $x_v \in \mathbb{R}^+$ give the concentrations of variables v. The vector x is called the quantitative state of G. The previous equations define the rate of change of each concentration x_v as the difference of the synthesis rate $f_v(x)$ and the degradation rate $\lambda_v x_v$ of v. The function f_v expresses how the synthesis rate of v depends on the concentrations x_u of its regulators $u \in G^-(v)$. It can be defined as,

$$f_v(x) = k_v +^\alpha \sum_{u \in G^-(v)} k_{uv} \; s^{\alpha_{uv}}(x_u, \theta_{uv}),$$

where:

- $k_v \in \mathbb{R}^+$ and $k_{uv} \in \mathbb{R}^{+*}$ are kinetic parameters,
- the function $s^{\alpha_{uv}}$ gives the effect of a regulator u and its target v. This function is usually a sigmoid depending on the sign α_{uv} and on the quantitative threshold $\theta_{uv} \in \mathbb{R}^{+*}$ of the interaction.

Since the qualitative thresholds t of G give the order of the continuous thresholds θ (see the section 5.2.1), it is required that for all targets $v' \in G^+(u)$ of u different than v, $\theta_{uv} < \theta_{uv'}$ if $t_{uv} < t_{uv'}$ and $\theta_{uv} > \theta_{uv'}$ if $t_{uv} > t_{uv'}$. By denoting θ_u^l the threshold(s) θ_{uv} such that $v \in G^+(u)$ and $t_{uv} = l$ we then have:

$$\theta_u^1 < \theta_u^2 < \ldots < \theta_u^l < \theta_u^{b_u}$$

The sigmoidal function $s^{\alpha_{uv}}$ is often approximated by a step function in order to make possible the analytical analysis of the system. $s^{\alpha_{uv}}$ is then defined as a Boolean function which indicates if u is or not a resource of v:

$$s^+(x_u, \theta_{uv}) = \begin{cases} 1, & \text{if } x_u > \theta_{uv} \\ 0, & \text{if } x_u < \theta_{uv} \end{cases} \quad \text{and} \quad s^-(x_u, \theta_{uv}) = 1 - s^+(x_u, \theta_{uv})$$

Notice that these functions are not defined for $x_u = \theta_{uv}$ and that the system becomes a piecewise linear equation system. Figure 5-6 gives an example of an UDS of the biological regulatory graph of figure 5-6.

$$\begin{cases} \dot{x}_u &= 20 + 35 \times s^-(x_v, 10) + 40 \times s^+(x_u, 20) - 5 \times x_u \\ \dot{x}_v &= 25 \times s^+(x_u, 16) - 2 \times x_v \end{cases}$$

Figure 5-6. Example of UDS of the biological regulatory graph of figure 5-3. Parameters are: $k_u = 20$, $k_{vu} = 35$, $k_{uu} = 40$, $\theta_{vu} = 10$, $\theta_{uu} = 20$, $\lambda_u = 5$ for the first equation and $k_v = 0$, $k_{uv} = 25$, $\theta_{uv} = 16$, and $\lambda_v = 2$ for the second. Notice that $\theta_{uv} < \theta_{uu}$ since $1 = t_{uv} < t_{uu} = 2$.

5.3.2 Discretization map and domains

Since the step functions $s^{\alpha_{uv}}$ are not defined for $x_u = \theta_{uv}$, the differential equation system is not defined for the states x for which at least one component x_v equals a threshold $\theta_{vv'}$, $v' \in G^+(v)$. Such states are called *singular states*. Consequently, the properties of the system can be analysed in the $|V|$-dimensional phase space Ω defined by

$$\Omega = \prod_{v \in V} \Omega_v \quad \text{with} \quad \Omega_v = \mathbb{R}^+ \setminus \{\theta_{vv'} \mid v' \in G^+(v)\} \text{ for all } v \in V.$$

Ω corresponds to the set of *regular states*. We are now in position to define the *discretization map* $d : \Omega \to Q$ by $d(x) = (d_v(x_v))_{v \in V}$ with, for every $v \in V$, $d_v : \Omega_v \to Q_v$ defined by

$$d_v(x_v) = |\{\theta_{vv'} \mid v' \in G^+(v) \text{ and } \theta_{vv'} < x_v\}|.$$

This discretization map gives directly the cardinal of the set of thresholds less than the concentration of v. If $d_v(x_v) = l$, then x_v is greater than the l smallest thresholds and less than others. For all $v \in G^+(u)$ we have $x_u > \theta_{uv} \Rightarrow d_u(x_u) \geq t_{uv}$ and $x_u < \theta_{uv} \Rightarrow d_u(x_u) < t_{uv}$. Consequently, for all state $x \in \Omega$:

$$s^{\alpha_{uv}}(x_u, \theta_{uv}) = 1 \Rightarrow u \in \omega_v(d(x)).$$

Then, for all $x \in \Omega$, f_v can be rewritten as

$$f_v(x) = k_v + \sum_{u \in \omega_v(d(x))} k_{uv}. \tag{2}$$

The infinite set of continuous states whose discretization gives $q \in Q$ is an hyper-rectangular region $D(q)$ of Ω, called *domain*, defined by:

$$D(q) = \prod_{v \in V} D_v(q_v) \quad \text{with} \quad D_v(q_v) = \{x_v \in \Omega_v \mid d_v(x_v) = q_v\} \text{ for all } v \in V.$$

A domain $D(q)$ is bounded by hyperplanes corresponding to thresholds: for all variable v, if $q_v > 0$ then $\theta_v^{q_v}$ is the lower bound of $D_v(q_v)$ and if $q_v < b_v$ then $\theta_v^{q_v+1}$ is the upper bound of $D_v(q_v)$ (see the figure 5-7).

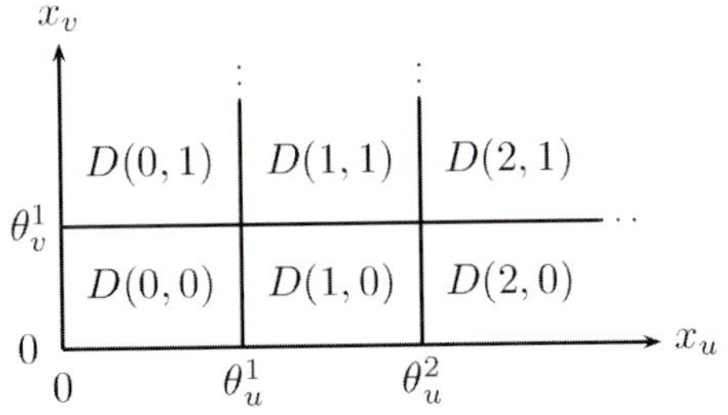

Figure 5-7. Domains of the phase space Ω of the UDS of figure 5-6 ($\theta_u^1 = \theta_{uv}$ and $\theta_u^2 = \theta_{uv}$).

5.3.3 Dynamics of differential equation systems

In a domain $D(q)$, each function f_v reduces to the constant $k_v + \sum_{u \in \omega_v(q)} k_{uv}$ (see equation 1). The system thus simplifies to a linear and uncoupled differential equation system whose solution in $D(q)$, starting at $x^0 \in D(q)$, is given by

$$x_v(t) = \kappa_v(x^0) - (\kappa_v(x^0) - x_v^0)\, e^{-\lambda_v t}, \quad \text{for all } v \in V,$$

with $\kappa_v(x) = f_v(x)/\lambda_v$ for all $x \in \Omega$. Function κ_v is also reduced to a constant $(k_v + \sum_{u \in \omega_v(q)} k_{uv})/\lambda_v$ in $D(q)$, denoted $\kappa_v(q)$ by abuse of notation. The state $\kappa(q) = (\kappa_v(q))_{v \in V}$ acts as an attractor in $D(q)$. Indeed, it is easy to verify that in $D(q)$, $x(t)$ has the following properties:

1. if $\kappa_v(q) \in D_v(q_v)$ then, $x_v(t)$ monotonically converges from x_v^0 towards $\kappa_v(q)$ and reaches $\kappa_v(q)$ in infinite time. Thus, if $\kappa(q) \in D(q)$, $x(t)$ does not leave $D(q)$ and the state $\kappa(q)$ is the unique stable steady state in $D(q)$.

2. if $\kappa_v(q) \notin D_v(q_v)$, then, if $x_v^0 < \kappa_v(q)$ (resp. $x_v^0 > \kappa_v(q)$), $x_v(t)$ monotonically increases (resp. decreases) from x_v^0 until to reach the threshold value $\theta_v^{q_v+1}$ (resp. $\theta_v^{q_v}$). The threshold is reached in a finite time iff $\kappa_v(q)$ is different from it. Throughout this section, we suppose that the parameters k and λ are taken such that $\kappa(q) \in \Omega$ for all $q \in Q$. Consequently, if $\kappa(q) \notin D(q)$, $x(t)$ leaves in a finite time $D(q)$ by reaching a threshold hyperplane.

If $\kappa_v(q) \notin D_v(q_v)$, $x_v(t)$ reaches its corresponding threshold at time t_v given by

$$t_v = -\frac{1}{\lambda_v} \ln \left(\frac{\kappa_v(q) - \theta_v^{q_v+\alpha}}{\kappa_v(q) - x_v^0} \right)$$

with $\alpha = 1$ if $x_v^0 < \kappa_v(q)$ and $\alpha = 0$ if $x_v^0 > \kappa_v(q)$. If at least two components $x_u(t)$ and $x_v(t)$ reach their thresholds simultaneously, one can deduce that x^0 belongs to an at most $(|V|-1)$-dimensional surface of zero Lebesgue measure in $D(q)$. Therefore, we do not consider this case and reason now as for almost every $x^0 \in D(q)$.

Suppose that t_v is the smallest value in $\{t_u \mid \kappa_u(q) \notin D_u(q_u)\}$, in other words, suppose that v is the variable whose concentration first leaves the domain. The component $x_v(t)$ reaches $\theta_v^{q_v+\alpha}$ at the singular state x^1 given by

$$x_v^1 = \theta_v^{q_v+\alpha} \quad \text{and} \quad x_u^1 = x_u(t_v) \quad \text{for all } u \neq v.$$

At this time, the trajectory exits from the domain $D(q)$ and enters into $D(q')$ defined by:

- $D_u(q'_u) = D_u(q_u)$ for all $u \neq v$, since only v reached its threshold,
- $D_v(q'_v) = D_v(q_v+1)$ if $\alpha = 1$ and $D_v(q'_v) = D_v(q_v-1)$ if $\alpha = 0$.

But at the singular state x^1, the differential equation system is not defined as well as $\kappa_v(x^1)$. The linear differential equation system of $D(q')$ is then extended by continuity to the hyperplane $x_v = \theta_v^{q_v+\alpha}$. Thus $\kappa_v(x^1)$ is defined, and the trajectory is extended with the solution of the differential system of $D(q')$ from the new starting point x^1.

5.3.4 Coherent discrete and differential modeling

Let $M(G)$ be a model of a biological regulatory graph $G=(V,E)$. The UDSs of G such that:

$$d_v(\kappa_v(q)) = K_{v,\omega_v(q)} \quad \text{for all } v \in V \text{ and } q \in Q$$

are called *underlying differential systems of M(G)*. A model $M(G)$ has UDSs if and only if it satisfies the Snoussi's constraints (equation 1) since we have:

$$d_v\left(\left(k_v + \sum_{u \in \omega} k_{uv}\right)/\lambda_v\right) = K_{v,\omega}, \quad \text{for all } v \in V \text{ and } \omega \subseteq G^-(v)$$

and thus $\omega \subseteq \omega'$ implies $\sum_{u \in \omega} k_{uv} \leq \sum_{u \in \omega'} k_{uv}$ which implies $K_{v,\omega} \leq K_{v,\omega'}$. For example the UDS of figure 5-6 is an UDS of the model described in Table 5-1. The following propositions show the coherence between the asynchronous dynamics of $M(G)$ and the dynamics of its UDSs.

Proposition 1
- *If there is an UDS of M(G) such that $x \in D(q)$ is a stable steady state, then q is a stable state of the asynchronous state graph S of M(G).*
- *Conversely, if q is a stable state of S then, for all UDSs of M(G), there is a stable steady state in the domain D(q).*

Proof. A state $x \in D(q)$ is a stable steady state iff $x_v = \kappa_v(q)$ for all $v \in V$. That implies $d_v(x_v) = d_v(\kappa_v(q)) \Rightarrow q_v = K_{v,\omega_v(q)}$ for all $v \in V$ and thus, q is a stable state of S. Conversely, if $q \in Q$ is a stable state, then $q_v = K_{v,\omega_v(q)} = d_v(\kappa_v(q))$ for all $v \in V$. Thus, $\kappa_v(q) \in D_v(q_v)$ for all $v \in V$ and consequently, $\kappa(q) \in D(q)$ is a stable steady state.

We define now the *boundary* of a domain as the set of singular states whose distance to the domain is null.

Proposition 2

- *If there is an UDS of M(G) for which there is a trajectory starting in D(q) which reaches directly from D(q) the hyperplane separating D(q) and an adjacent domain D(q'), then q → q' is a transition of the asynchronous state graph S of M(G).*
- *Conversely, there exist UDSs of M(G) such that, for each successor q' of q in S, there is a trajectory starting in D(q) which reaches directly from D(q) the hyperplane separating D(q) and D(q').*

Proof. We have seen in section 5.3.3 that if a trajectory starting at $x^0 \in D(q)$ reaches the hyperplane separating $D(q)$ and an adjacent domain $D(q')$, then there is a unique variable $v \in V$ such that $q'_v \neq q_v$ and we have $q'_v = q_v + 1$ if $x_v^0 < \kappa_v(q)$ or $q'_v = q_v - 1$ if $x_v^0 > \kappa_v(q)$. Moreover, $\kappa_v(q) \notin D_v(q_v)$ thus $x_v^0 < \kappa_v(q)$ iff $d_v(x_v^0) < d_v(\kappa_v(q))$ which is equivalent to $q_v < K_{v,\omega_v(q)}$. Similarly $x_v^0 > \kappa_v(q)$ iff $d_v(x_v^0) > d_v(\kappa_v(q))$ which is equivalent to $q_v > K_{v,\omega_v(q)}$. According to definition 5, $q \to q'$ is a transition of S.

Now, we prove the second part of the proposition. Consider the UDSs of $M(G)$ such that $\lambda_u = \lambda$ for all $u \in V$ and an initial state $x^0 \in D(q)$. The trajectory starting at x^0 describes the part of the segment connecting x^0 to $\kappa(q)$ which belongs to $D(q)$.

Let q' be a successor of q in S. We have $\kappa(q) \notin D(q)$. Let us choose a point x^1 of the boundary of $D(q)$ belonging to the hyperplane separating $D(q)$ from the domain $D(q')$ and whose only one component equals a threshold. The trajectories starting at a point of the line connecting x^1 and $\kappa(q)$ which belongs to $D(q)$, reach x^1.   □

We deduce from the previous propositions that all the regular stable steady states of an UDS of $M(G)$ are represented in its asynchronous state graph S. Moreover if a trajectory of an UDS of $M(G)$ passes successively through the domains $D(q^0)$, $D(q^1)$, ..., $D(q^n)$ then $q^0 \to q^1 \to ... \to q^n$ is a path of S. But if $q^0 \to q^1 \to ... \to q^n$ is a path of S, it does not mean that there is a trajectory passing successively through the domains $D(q^0)$, $D(q^1)$, ..., $D(q^n)$. Using the terminology of [14], the qualitative modelling is said sound. A graphical comparison between the asynchronous dynamics of a model and a trajectory of one of its UDS is given in figure 5-8.

Any UDS of a biological regulatory graph G is an UDS of a model of G satisfying the Snoussi's constraints. Thus the trajectories of the infinite set of UDSs of G are summarized by a finite set of asynchronous state graphs (for the biological regulatory graph of figure 5-3, we have 42 different state graphs deduced from the 60 different models satisfying the Snoussi's constraints).

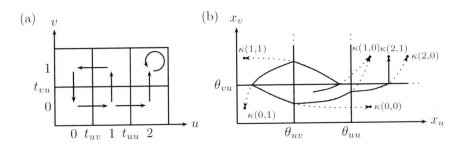

Figure 5-8. (a) The asynchronous state graph of the model *M(G)* of Table 5-1. (b) A trajectory of an UDS of *M(G)*. The dotted arrows represent the extensions of solutions towards the attractors.

5.3.5 Feedback circuit functionality

Feedback circuits play a major role for the dynamics of systems since they can generate multi-stationarity or homeostasis. A positive (resp. negative) circuit is said *functional* if it generates multi-stationarity (resp. homeostasis). The functionality of circuits is strongly related to the the stationarity of particular singular states and to discontinuities of the UDS. To deal with them, we first introduce the differential inclusion systems.

5.3.5.1 Differential inclusion systems

To deal with ordinary differential equation systems with discontinuous right-hand sides, Filippov [9] proposed to extend them to systems of differential inclusions. For the regulatory networks, the UDSs can be extended to the following differential inclusions systems:

$$x_v \in H_v(x), \text{ for all } v \in V,$$
$$(3)$$

where H_v is a set-valued function, defined as follow:

- for all regular state x, $H_v(x) = \{f_v(x) - \lambda_v x_v\}$. For all $x \in D(q)$, since $f_v(x)/\lambda_v = \kappa_v(q)$, $H_v(x)$ can be rewritten as $H_v(x) = \{\lambda_v(\kappa_v(q) - x_v)\}$.

- for all singular state x,

$$H_v(x) = \overline{co} \ (\{\lambda_v(\kappa_v(q) - x_v) \mid q \in N(x)\}).$$

- where $\overline{co}(E)$ designs the smallest closed convex set of a set E which is the intersection of all closed convex sets containing E, and where $N(x)$ is the set of qualitative states which correspond to domains whose boundary contains x:

$$N(x)=\left\{ q\in Q \;\middle|\; \forall u\in V, q_u = \begin{cases} d_u(x_u), & \text{if } x_u\in\Omega_u \\ t_{uv}-1 \text{ or } t_{uv} \text{ if } x_u=\theta_{uv} \text{ with } v\in G^+(u) \end{cases} \right\}.$$

- Obviously we have

$$H_v(x)=\left[\min_{q\in N(x)} \lambda_v\big(\kappa_v(q)-x_v\big) , \; \max_{q\in N(x)} \lambda_v\big(\kappa_v(q)-x_v\big) \right]$$

Consider the example of figure 5-7. For x such that $x_u = \theta_u^2 = \theta_{uu}$ and $x_v > \theta_{uv}^2$, we have $N(x) = \{(1,1), (2,1)\}$ and for these states, $\omega_u(1,1) = \{\}$ is included in $\omega_u(2,1) = \{u\}$ and $\omega_v(1,1) = \omega_v(2,1) = \{u\}$. We deduce that $H_u(x) = [\lambda_u(\kappa_u(1,1) - x_u), \lambda_u(\kappa_u(2,1) - x_u)]$ and $H_v(x) = \{\lambda_v(\kappa_v(1,1) - x_v)\}$. Intuitively, at the singular state x, the regulation $u \to v$ is clearly defined: $s^+(x_u,\theta_{uv}) = 1$. This is why the set $H_v(x)$ of the possible derivatives of x_v is single-valued. However, as $x_u = \theta_{uu}$ the self regulation of u remains undefined and $H_u(x)$ is *a priori* not single-valued: the derivative of x_u is comprised between the derivatives obtained with $s^+(\theta_{uu},\theta_{uu}) = 0$ and $s^+(\theta_{uu},\theta_{uu}) = 1$.

An absolutely continuous function $x(t)$ is solution of the system (3) in the sense of Filippov if $\dot{x}_v(t) \in H_v(x(t))$ for all $v\in V$ and for almost all $t \geq 0$. The qualification ``for almost all $t \geq 0$" means that the set time-points for which the condition does not holds if of measure 0. In particular, the condition is not satisfied at time-points when the solution arrives or leaves a threshold hyperplane.

We do not analyse the solutions in the sense of Filippov in this section (see [7, 11] for a detailed analysis), but the previous formalism will be useful for analysis of the steadiness of singular states.

5.3.5.2   Steadiness of singular states

It is not surprising that a state x, regular or singular, is an equilibrium point (in the sense that there is a solution $x(t)$ such that $x(t) = x$ for all $t \geq 0$) when $0\in H_v(x)$ for all $v\in V$. For a regular state $x\in D(q)$, we have, as for differential equation systems:

$$0\in H_v(x) \;\Rightarrow\; 0\in\{ \lambda_v(\kappa_v(q) - x_v) \} \;\Rightarrow\; x_v = \kappa_v(q).$$

In this case, x is a regular stable steady state. For a singular state, the inclusion can be written as an inequality:

$$0\in H_v(x) \quad\Leftrightarrow\quad \min_{q\in N(x)} \lambda_v\big(\kappa_v(q)-x_v\big) \leq 0 \leq \max_{q\in N(x)} \lambda_v\big(\kappa_v(q)-x_v\big)$$

$$\Leftrightarrow\quad \min_{q\in N(x)} \kappa_v(q) \leq x_v \leq \max_{q\in N(x)} \kappa_v(q)$$

and if $x_v \notin \Omega_v$ the inequality becomes strict:

$$0 \in H_v(x) \quad \Leftrightarrow \quad \min_{q \in N(x)} \kappa_v(q) < x_v < \max_{q \in N(x)} \kappa_v(q)$$

because $\kappa_v(q) \in \Omega_v$ for all $q \in N(x)$. Among all singular equilibrium points, those for which we have $\min_{q \in N(x)} \kappa_v(q) = x_v = \max_{q \in N(x)} \kappa_v(q)$ for $x_v \in \Omega_v$, are *singular steady states* [7, 24]. Figure 5-9 shows a graphical representation of the conditions for the steadiness of singular states.

Proposition 3 *Let x be a singular state and v a variable. If for all $u \in G^-(v)$ $x_u \neq \theta_{uv}$, then $\kappa_v(q)$ is constant for all $q \in N(x)$.*

Proof. For all $u \in G^-(v)$ we have $x_u \in \Omega_u$ or $x_u = \theta_{uv'} \neq \theta_{uv}$ with $v' \in G^+(u)$. In the first case, it is evident that $q_u = q'_u$ for all q and q' in $N(x)$. In the second case, for all q and q' in $N(x)$, q_u and q'_u belong to $\{t_{uv'}-1, t_{uv'}\}$ and $t_{uv'} \neq t_{uv}$. Then q_u and q'_u are on the same side of t_{uv}. Consequently, for all q and q' in $N(x)$ we have $\omega_v(q) = \omega_v(q')$ which implies $\kappa_v(q) = \kappa_v(q')$. $\square$

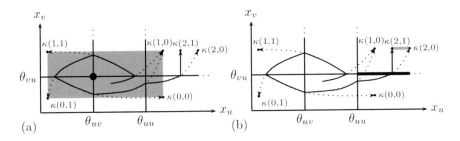

Figure 5-9. Equilibrium points and their steadiness. Grey regions, a rectangle in (a) and a segment in (b), correspond to the Cartesian product $\Psi(x) = [\min_{q \in N(x)} \kappa_u(q), \max_{q \in N(x)} \kappa_u(q)] \times [\min_{q \in N(x)} \kappa_v(q), \max_{q \in N(x)} \kappa_v(q)]$ for a singular state x. In (a) $x = (\theta_{uv}, \theta_{vu})$ is an equilibrium point ($x \in \Psi(x)$) and since all variables are singular, it is steady. In (b) the singular state is such that $x_u > \theta_{uu}$ and $x_v = \theta_{vu}$ and it is not an equilibrium point ($x \notin \Psi(x)$).

5.3.5.3   Circuit characteristic states

Definition 6 [Circuit] *Let $G = (V,E)$ be a biological regulatory graph. A circuit of G is a finite sequence of distinct elements of V, denoted $C = v_1, v_2, ..., v_n$, such that $v_n \to v_1 \in E$ and $v_i \to v_{i+1} \in E$ for all $i \in \{1, ..., n-1\}$.*

In the sequel, $\{C\}$ denotes the set of variables of a circuit C and $i+1$ (resp. $i-1$) is always computed modulo n: v_{i+1} (resp. v_{i-1}) denotes the successor (resp. predecessor) of v_i in C. Two circuits C and C' are disjointed if they have no variable in common. In a pedagogical objective, we focus here on the properties of a single circuit, but all results can be extended to a union of disjointed circuits [24]. Moreover we take into consideration only regulatory graphs where for any variable the out-thresholds are distinct ($b_v = |G^+(v)|$, $\forall\ v \in V$).

A singular state x is said *characteristic* of a circuit $C = v_1, v_2, ..., v_n$ if the concentration x_{v_i} of each variable v_i of the circuit is equal to the threshold $\theta_{v_i v_{i+1}}$ and if the concentrations of other variables are regular: $x_u \in \Omega_u$, $\forall\ u \notin \{C\}$.

Proposition 4 *A singular steady state is a characteristic state of a circuit of the biological regulatory graph G.*

Proof. Let x be a singular state and $S = \{v \mid x_v \notin \Omega_v\}$ be the set of variables equal to a threshold at the state x. If x is steady, we have for all $v \in S$:

$$\min_{q \in N(x)}\ \kappa_V(q) < x_v < \max_{q \in N(x)}\ \kappa_V(q).$$

According to the proposition 3, if for all $u \in G^-(v)$ we have $x_u \neq \theta_{uv}$ then $\min_{q \in N(x)} \kappa_v(q) = \max_{q \in N(x)} \kappa_v(q)$ and x is not steady. Thus v has at least one predecessor u such that $x_u = \theta_{uv}$, which implies that $u \in S$. Moreover, because $\theta_{uv'} \neq \theta_{uv}$ for all $v' \in G^+(u)$, the successor v of u is the only one such that $x_u = \theta_{uv}$. Each variable v of S has then a unique predecessor u in S such that $x_u = \theta_{uv}$.

□

Now, we prove that all the steady singular states can be identified in the qualitative modelling.

Proposition 5 *Let G be a biological regulatory graph containing a circuit $C = v_1, ..., v_n$. Consider a UDS of a model M(G) and a characteristic state x of C. Let $q \in N(x)$. If x is steady, then M(G) is such that*

$$\begin{cases} K_{v, \omega_v}(q) = q_v \\[2mm] v \notin \{C\} \end{cases} \qquad \text{for all}$$

$$K_{v_i, \omega_{v_i}}(q) \setminus \{v_{i-1}\} \ < \ t_{v_i v_{i+1}} \ \leq \ K_{v_i, \omega_{v_i}}(q) \cup \{v_{i-1}\}$$
$$\text{for all}$$

$$i \in \{1, ..., n\}$$

Proof. Let $v \notin \{C\}$. Since x is characteristic of C, we have $x_v \in \Omega_v$. If x is steady, then $\min_{q \in N(x)} \kappa_v(q) = x_v = \max_{q \in N(x)} \kappa_v(q)$. That means that $\kappa_v(q)$ is constant for all $q \in N(x)$ and we have $d_v(x_v) = d_v(\kappa_v(q))$ which is equivalent to $q_v = K_{v, \omega_v}(q)$.

Let $v_i \in \{C\}$. As x is characteristic of C, v_{i-1} is the unique predecessor of v_i such that $x_{v_{i-1}} = \theta_{v_{i-1} v_i}$. Thus, $\omega_{v_i}(q) \setminus \omega_{v_i}(q')$ equals $\{\}$ or $\{v_{i-1}\}$ for all q and q' in $N(x)$. Moreover there is at least one state $q \in N(x)$ such that $q_{v_{i-1}} = t_{v_{i-1} v_i}$ and another one such that $q_{v_{i-1}} = t_{v_{i-1}v_i} - 1$. Thus, there is a state $q^+ \in N(x)$ such that $v_{i-1} \in \omega_{v_i}(q^+)$ and a state q^- with $v_{i-1} \notin \omega_{v_i}(q^-)$. We deduce that for all $q \in N(x)$, $\omega_v(q) \cup \{v_{i-1}\} = \omega_{v_i}(q^+)$ and $\omega_v(q) \setminus \{v_{i-1}\} = \omega_{v_i}(q^-)$. So $\max_{q \in N(x)} \kappa_{v_i}(q) = \kappa_{v_i}(q^+)$ and $\min_{q \in N(x)} \kappa_{v_i}(q) = \kappa_{v_i}(q^-)$. Since $x_{v_i} = \theta_{v_i v_{i+1}}$, if x is steady, we have for all $q \in N(x)$:

$$\kappa_{v_i}(q^-) < \theta_{v_i v_{i+1}} < \kappa_{v_i}(q^+) \quad \Leftrightarrow \quad d_{v_i}(\kappa_{v_i}(q^-)) < \theta_{v_i v_{i+1}} < d_{v_i}(\kappa_{v_i}(q^+))$$

$$\Leftrightarrow \quad K_{v_i, \omega_{v_i}(q^-)} < t_{v_i v_{i+1}} \le K_{v_i, \omega_{v_i}(q^+)}$$

$$\Leftrightarrow \quad K_{v_i, \omega_{v_i}(q) \setminus \{v_{i-1}\}} < t_{v_i v_{i+1}} \le K_{v_i, \omega_{v_i}(q) \cup \{v_{i-1}\}}$$

Definition 7 [Quasi-characteristic qualitative states]   *Let $G=(V,E)$ be a biological regulatory graph containing a circuit $C = v_1, \ldots, v_n$. A state $q \in Q$ is quasi-characteristic of C if $q_{v_i} = t_{v_i v_{i+1}}$ for all $v_i \in \{C\}$.*

The quasi-characteristic states are useful to locate the singular characteristic states of the UDS.

Proposition 6   *Let G be a biological regulatory graph containing a circuit $C = v_1, \ldots, v_n$ and a quasi-qualitative characteristic state q of C. If a model $M(G)$ satisfies the Snoussi's constraints and if*

$$\begin{cases} K_{v, \omega_v(q)} = q & \text{for all} \\ v \notin \{C\} \end{cases} \tag{4}$$

$$K_{v_i, \omega_{v_i}(q) \setminus \{v_{i-1}\}} < t_{v_i v_{i+1}} \le K_{v_i, \omega_{v_i}(q) \cup \{v_{i-1}\}} \quad \text{for all}$$

$$i \in \{1, \ldots, n\}$$

then, for all the UDSs of $M(G)$, there exists a unique steady characteristic state x of C such that $d_u(x_u) = q_u$ for all $u \notin \{C\}$.

As the proof is quite similar to the previous one, it is omitted.

The previous proposition makes easy the determination of all steady singular states underlying of a qualitative model. Let us consider for instance the model $M(G)$ of Table 5-1. The corresponding biological regulatory graph G (Figure 5-3-(a)) contains two circuits, $C^1 = u,v$ and $C^2 = u$. The unique quasi-characteristic state of C^1 is (t_{uv}, t_{vu}). It satisfies

$$K_{u, \omega_u(t_{uv}, t_{vu}) \setminus \{v\}} < t_{uv} = 1 \le K_{u, \omega_u(t_{uv}, t_{vu}) \cup \{v\}} \quad \text{and} \quad K_{v, \{\}} < t_{vu} = 1$$

$$\le K_{v, \{u\}}.$$

Indeed the first inequality is verified because $\omega_u(t_{uv}, t_{vu}) = \{\}$, $K_{u,\{\}} = 0$ and $K_{u,\{v\}} = 2$, the second is also verified since $K_{v,\{\}} = 0$ and $K_{v,\{u\}} = 1$. Consequently, the characteristic state $(\theta_{uv}, \theta_{uv})$ is steady in all the UDSs of $M(G)$.

For circuit C^2, there are two quasi-characteristic states: $(t_{uu}, 0)$ and $(t_{uu}, 1)$.

- The first one, $(t_{uu}, 0)$, does not satisfy

  $$K_{u,\omega_u(t_{uu},0)\backslash\{u\}} < t_{uu} = 2 \leq K_{u,\omega_u(t_{uv},0) \cup \{u\}} \quad \text{and} \quad K_{v,\omega_v(t_{uu},0)} = 0.$$

  since $\omega_v(2,1) = \{u\}$ and $K_{v,\{u\}} = 1$. Thus there is not any steady characteristic state of C^2 such that $x_v < \theta_{vu}$.

- The second quasi-characteristic state, $(t_{uu}, 1)$, satisfies

  $$K_{u,\omega_u(t_{uu},1)\backslash\{u\}} < t_{uu} = 2 \leq K_{u,\omega_u(t_{uv},1) \cup \{u\}} \quad \text{and} \quad K_{v,\omega_v(t_{uu},1)} = 1.$$

  since $\omega_v(2,1) = \{u\}$, $K_{v,\{u\}} = 1$, $\omega_u(2,1) = \{u\}$, $K_{u,\{\}} = 0$ and $K_{u,\{u\}} = 2$. For all UDSs of $M(G)$ there is a unique steady characteristic state x of C^2 such that $x_v > \theta_{vu}$.

The detected singular states are represented in the asynchronous state graph of $M(G)$ in figure 5-10.

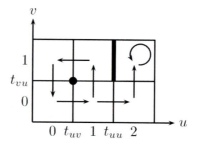
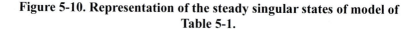

**Figure 5-10. Representation of the steady singular states of model of
Table 5-1.**

5.3.5.4   Circuit functionality

Each variable of a feedback circuit has an influence on its target but also an indirect effect on all following variables including itself. A circuit is said positive (resp. negative) if each variable has a positive (resp. negative) influence on itself. The sign of a circuit is determined by the number of inhibitions: if it is odd, the circuit is negative and otherwise, the circuit is positive. Negative and positive circuits have different typical behaviours.

- In a negative circuit, a high level of a variable tends to make decrease itself and conversely. Thus the circuit makes the level of each variable to tend to (or oscillate around) an equilibrium concentration. It generates stable oscillation behaviour corresponding to homeostasis in biology.
- In a positive circuit, a high (resp. low) level of a variable tends to make it increase (resp. decrease). Thus each variable stays either at a low or high concentration and the positive circuit generates multi-stationarity corresponding to differentiation in biology.

A circuit which presents a typical behaviour is said functional. Several authors have proved that at least one positive circuit is necessary to generate multi-stationarity whereas at least one negative circuit is necessary to obtain a stable oscillatory behaviour [5, 6, 18, 23, 25]. Snoussi and Thomas realized that when a characteristic state is steady, the corresponding circuit is functional [24]. In the qualitative formalism, the circuit functionality is then defined as follow.

Definition 8 [Functional circuit]   *Let M(G) be a model of a biological regulatory graph G containing a circuit C. If there is a quasi-characteristic state q of C satisfying the constraint (4) then C is functional.*

We deduce from the proposition 6 that if a circuit C is functional, there is, for all underlying differential systems, a steady characteristic state x of C such that $x_u = d_u(q_u)$ for all $u \notin \{C\}$. In the model of Table 5-1, both circuits $u \rightarrow u$ and $u \rightarrow v \rightarrow u$ are functional. As a result, multi-stationarity and homeostasis are present in the corresponding asynchronous state graph (figure 5-10).

Summing up, homeostasis and/or multi-stationarity are dynamical properties almost always present in biological systems. Circuit functionality is then useful for modelling such systems. For example, it has been used to model immunity control in lambda phage [26], pattern formation during the embryonic development of *Drosophilae* [19, 20] and flower morphogenesis in *Arabidopsis thaliana* [16].

5.4 Formal methods

To study the behaviour of the genetic regulatory network, the ordinary differential equation systems are well adapted if all the parameters are well known. Unfortunately they are most often unknown and are difficultly measurable *in vivo*. The discrete approach of Thomas and co-workers simplifies the problem of determining the suitable parameters since the number of possible models is finite. Indeed finding suitable classes of those parameters constitutes a major issue of the modelling activity. Even if the Snoussi's constraints on parameters are used, the number of remaining models is too large to analyse them by hand. Then biological knowledge or hypotheses on the behaviour of the system can be used as an indirect criterion to constrain the parameters. For example homeostasis (resp. multi-stationarity) is experimentally observable and it indicates that a negative (resp. positive) feedback circuit is functional, this functionality leading to some constraints on the parameters (see section 5.3).

To go further, conditions of multi-stationarity and homeostasis can be reinforced by introducing other conditions on the dynamics of the system. The available knowledge on the evolution of the system, as temporal properties, can be taken into consideration for constraining the values of parameters. Among all suitable models only a part of them are coherent with these temporal properties. Since numerous models have to be checked against those properties, a formal language is needed to perform automatically these checkings.

5.4.1 Temporal logic

The properties as the deadlock can be easily checked by exploring the transition system, called *asynchronous state graph* in section 5.2. For more complex properties on the dynamics of the system it is necessary to use a well adapted formal language: a temporal language which allows the specification of properties along the execution paths of the transition system. The step of the specification of the properties can then be distinguished from the specification of the system since it is not necessary to know the dynamic structure of the system to be checked for specifying the properties.

Expressing temporal properties on a transition system needs to define the atomic propositions which depends of the considered regulatory graph $G=(V,E)$. Generally the set of atomic propositions is denoted by AP. The subset of AP containing all the atomic propositions which are true in a state q, is given by the labelling function $\mathcal{L}$:

$$\mathcal{L}(q) = \{ (v = q_v) \mid v \in V \}$$

where $(v = q_v)$ signifies that the variable v has the concentration level q_v. The pair composed of a transition system and a labelling function is called a Kripke structure.

Execution traces of the transition system model implicitly a discrete time: if an execution passes from the state s_0 to s_1, the instant associated to the state s_1 follows the one corresponding to the state s_0. The temporal logics allow one to specify dynamical properties referring to this discrete time [8]. The Linear Temporal Logic, LTL, is used to specify properties on an execution of the system. If the system is determinist, from any initial state there is a unique execution, LTL is appropriated to specify properties of the system. Nevertheless the qualitative behaviour of a biological regulatory network is represented by an asynchronous state graph, which is non determinist: the current state can have several possible futures. Since time has a tree structure, we prefer the Computation Tree Logic, CTL, in which it is possible to express properties of the form "*it is possible in the future that...*".

Definition 9 [Syntax of CTL] *A CTL formula on the set of atomic propositions AP is inductively defined by:*

- $\top, \bot$ *and any atomic proposition of AP are formulae*
- *if ϕ and ψ are formulae, then* $(\neg\phi)$, $(\phi\wedge\psi)$, $(\phi\vee\psi)$, $(\phi\Rightarrow\psi)$, $(\phi\Leftrightarrow\psi)$, $AX\phi$, $EX\phi$, $A[\phi U\psi]$, $E[\phi U\psi]$, $AG\phi$, $EG\phi$, $AF\phi$, $EF\phi$ *are formulae.*

The semantics of CTL is defined on the execution trees of the transition system which are completely defined by their initial state and the transition relation. The semantics is given by the definition of the satisfaction relation $s \models \phi$ meaning that the formula ϕ is satisfied on the execution tree starting at s.

Definition 10 [Semantics of CTL] *Let s_0 be a state. The semantics of CTL is defined inductively by:*

- $s_0 \models \top$ *and* $s_0 \not\models \bot$,
- $\forall p \in AP, s_0 \models p$ *iff* $p \in \mathcal{L}(s_0)$,
- $s_0 \models \neg\varphi$ *iff* $s_0 \not\models \varphi$,
- $s_0 \models \varphi_1 \wedge \varphi_2$ *iff* $s_0 \models \varphi_1$ *and* $s_0 \models \varphi_2$,
- $s_0 \models \varphi_1 \vee \varphi_2$ *iff* $s_0 \models \varphi_1$ *or* $s_0 \models \varphi_2$,
- $s_0 \models \varphi_1 \Rightarrow \varphi_2$ *iff* $s_0 \not\models \varphi_1$ *or* $s_0 \models \varphi_2$,
- $s_0 \models \varphi_1 \Leftrightarrow \varphi_2$ *iff* $s_0 \models (\varphi_1 \Rightarrow \varphi_2) \wedge (\varphi_2 \Rightarrow \varphi_1)$,
- $s_0 \models AX\varphi$ *iff for all successors s_1 of s_0, we have $s_1 \models \varphi$,*
- $s_0 \models EX\varphi$ *iff for any successor s_1 of s_0, we have $s_1 \models \varphi$,*
- $s_0 \models AG\varphi$ *iff for all paths $s_0, s_1...s_i...$, and for all s_i along the path we have $s_i \models \varphi$,*
- $s_0 \models EG\varphi$ *iff for a particular path $s_0, s_1...s_i...$ we have for all s_i along the path $s_i \models \varphi$,*
- $s_0 \models AF\varphi$ *iff for all paths $s_0, s_1...s_i...$, there exists s_i along the path such that $s_i \models \varphi$,*
- $s_0 \models EF\varphi$ *iff for a particular path $s_0, s_1...s_i...$, there exists s_i along the path such that $s_i \models \varphi$,*
- $s_0 \models A[\varphi_1 U \varphi_2]$ *iff for all paths $s_0, s_1...s_i...$, there exists s_i along the path such that $s_i \models \varphi_2$ and for each $j < i$ we have $s_j \models \varphi_1$,*
- $s_0 \models E[\varphi_1 U \varphi_2]$ *iff for a particular path $s_0, s_1...s_i...$, there exists s_i along the path such that $s_i \models \varphi_2$ and for each $j < i$ we have $s_j \models \varphi_1$.*

$\top$ is the always true formula; $\bot$ is the always false formula; a state s satisfies all the atomic formulae of $\mathcal{L}(s)$; $\neg$, $\wedge$, $\vee$, $\Rightarrow$, $\Leftrightarrow$ are the usual connectives (respectively *not, and, or, implication, equivalence*). All the temporal connectives are pairs of symbols: the first element is A or E followed by X, F, G or U whose meanings are given in the next table and illustrated in Figure 5-11.

A for **A**ll paths choices	X ne**X**t state
E for at least one path choice (**E**xist)	F some **F**uture state
	G all future states (**G**lobally)
	U **U**ntil

Consider the example of Figure 5-5(b) where variables are u and v. The atomic proposition are $AP=\{(u = 0),(u = 1),(u = 2),(v = 0),(v = 1)\}$. AX($v = 1$) means that in all next states accessible from the current state in the asynchronous state graph, the concentration level of v is 1. This formula is true iff the current state is $(1,1)$, $(2,0)$ or $(2,1)$. EG($\neg(u = 2)$) means that there exists at least one path starting from the current state where the concentration level of u is constantly strictly less than 2. In Figure 5-5(b), all states for which u is strictly less than 2 satisfy the formula. Then $\neg(u = 2) \Rightarrow$ EG($\neg(u = 2)$) is satisfied for all states. A[$(v = 1)$U$(v = 0)$] means that for any possible path from the current state there exists a future state where $v = 0$ and in between v remains equal to 1. Note that $(2,1)$ is the only state, which does not satisfy the formula. And so on for other temporal connectives.

It is now possible to translate a biological temporal property into a CTL formula. Classically a biological system can have several steady states corresponding to distinct phenotypes. Let us suppose that two distinct stable states, ss_1 and ss_2, are possible and that formulae ψ_1 and ψ_2 characterize the states ss_1 and ss_2 respectively. If the system is able to go from state s_0, characterized by the formula φ_0, either to state ss_1 or to state ss_2, these temporal properties can be translate into formulae:

$\psi_1 \Rightarrow$ AG ψ_1	stability of state ss_1
$\psi_2 \Rightarrow$ AG ψ_2	stability of state ss_2
$(\varphi_0 \Rightarrow$ EF$\psi_1) (\varphi_0 \Rightarrow$ EF$\psi_2)$	reachability of ss_1 and ss_2 from s_0

Such formulae are used in the concrete example of section 5.5 for expressing biological knowledge on the immunity control in bacteriophage lambda.

5.4.2 Model checking

The model checking is a verification method that proves automatically if a Kripke structure satisfies a temporal formula [13]. We briefly present the basic algorithm of model checking for a CTL formula. Since the connectives $\vee$, $\Rightarrow$ and $\Leftrightarrow$ can be rewritten in term of $\neg$ and $\wedge$, and since we have the following equivalence:

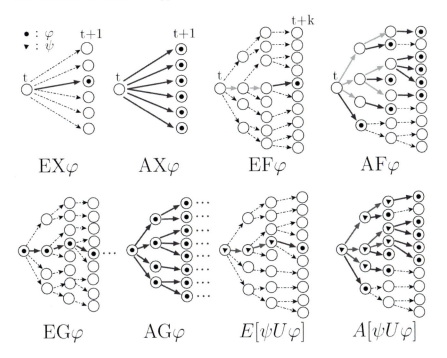

Figure 5-11. Semantics of temporal connectives of CTL.

$$AX\varphi \equiv \neg EX(\neg\varphi)$$
$$EG\varphi \equiv \neg AF(\neg\varphi)$$
$$EF\varphi \equiv E(\top\ U\ \varphi)$$
$$AG\varphi \equiv \neg EF(\neg\varphi)$$
$$A[\varphi_1U\varphi_2] \equiv \neg\,(\,E[\neg\varphi_2U(\neg\varphi_1\wedge\neg\varphi_2)] \vee EG(\neg\varphi_2)\,)$$

We consider in the sequel formulae containing only the connectives: $\neg$, $\wedge$, EX, AF and EU. Obviously any CTL formula can be transformed into a semantically equivalent CTL formula, which uses only those connectives.

The model checking for a CTL formula φ consists of labelling each state s of the transition system with sub-formulae of φ, which is satisfied at the state s. These sub-formulae are added to $\mathcal{L}(s)$ containing initially the atomic propositions true in s. Suppose that ψ is a sub-formula of φ and that states satisfying all the immediate sub-formulae of ψ have already been labelled. The labelling algorithm for ψ uses a case analysis to label states with ψ:

- if $\psi \in AP$, then the labelling is given directly by $\mathcal{L}(s)$
- if $\psi = p \wedge q$, then $\mathcal{L}(s) = \mathcal{L}(s) \cup \{p \wedge q\}$ for all s such that $p, q \in \mathcal{L}(s)$
- if $\psi = \neg p$, then $\mathcal{L}(s) = \mathcal{L}(s) \cup \{\neg p\}$ for all s such that $p \notin \mathcal{L}(s)$

- if $\psi = EXq$, then $L(s) = L(s) \cup \{EXq\}$ for all predecessors s of a state t such that $q \in L(t)$
- if $\psi = AFq$, then
    1. $L(s) = L(s) \cup \{AFq\}$ for all s such that $q \in L(s)$
    2. Repeat: $L(s) = L(s) \cup \{AFq\}$ for all states s such that all successors are labelled with AF q, until there is no change.
- if $\psi = E[qUr]$, then
    1. $L(s) = L(s) \cup \{E[qUr]\}$ for all s such that $r \in L(s)$,
    2. Repeat: $L(s) = L(s) \cup \{E[qUr]\}$ for all states s such that $q \in L(s)$ and which have a successor labelled with $E[qUr]$, until there is no change.

It can be proved that this labelling algorithm ends and that states are labelled with all sub-formulae of φ that they satisfy. Thus $s \models \varphi$ if the state s is labelled with φ. By extension if all states are labelled with φ, we say that the considered Kripke structure satisfies φ.

The model checking algorithm is linear with the size of the system and the size of the formula. Unfortunately, practical applications lead to transition systems with an enormous number of states, and the previous algorithm is often inefficient. To push back these limits, *symbolic* model checking [15] has been developed. It consists in computations on symbolic representation of subspaces of states.

To sketch the symbolic model checking, let us introduce the operator *Pre*. Let S be the set of states and x be a subset of S. $Pre(x)$ gives the set of states which have a successor in x. The set $sat(\varphi)$ of states satisfying φ can then be defined inductively:

- if $\varphi \in AP$, $sat(\varphi) = \{s \in S \mid \varphi \in L(s)\}$
- $sat(\neg\varphi) = S \setminus sat(\varphi)$
- $sat(\varphi \vee \psi) = sat(\varphi) \cup sat(\psi)$
- $sat(\varphi \wedge \psi) = sat(\varphi) \cap sat(\psi)$
- $sat(EX\varphi) = Pre(sat(\varphi))$
- $sat(AX\varphi) = S \setminus Pre(S \setminus sat(\varphi))$
- The connectives $AF\varphi$ and $E[\varphi_1 U\varphi_2]$ are more difficult to define. Let us remark that we have the following equivalence:

$$
\begin{aligned}
AF\ \varphi &\equiv \varphi \vee (AX\ (AF\varphi)) \\
E[\varphi_1 U\varphi_2] &\equiv \varphi_2 \vee (\varphi_1 \wedge EX(E[\varphi_1 U\varphi_2])\ ).
\end{aligned}
$$

Then $sat(AF\varphi)$ and $sat(E[\varphi_1 U\varphi_2])$ can be defined as the smallest fixed points of equations:

$$f_1(x) = sat(\varphi) \cup sat(\text{AX } x)$$
$$f_2(x) = sat(\varphi_2) \cup (sat(\varphi_1) \cap sat(\text{EX } x)).$$

Since functions f_1 and f_2 are monotone and that the set of states is finite, the iterative computation of the smallest fixed point ends.

The *Binary Decision Diagrams*, or BDD for short, are data structures allowing the representation of Boolean expressions in a very compact way. Then subsets of states can be coded with such Boolean expressions and necessary operations for computing *sat* can be defined on these structures. Numerous works detail utilization of BDDs for the verification of systems, see for example [13, 15].

5.4.3   A tool for the selection of models: SMBioNet

We have designed a software for a computer aided modelling based on the previous described formal methods [3]. This software, SMBioNet[1], helps the biologist and/or the modeller to verify systematically the coherence of models of a given biological system, and to select suitable models which satisfy the temporal properties extracted from knowledge or hypothesis. More precisely inputs of SMBioNet consist in:

*   a biological regulatory graph representing the interactions of the biological system and
*   a CTL formula expressing its known or hypothetical dynamical properties.

Then it generates all the models of the biological regulatory graph and gives as output those satisfying the CTL formula. For each generated model, SMBioNet calls the model checker NuSMV [4] and selects it if the formula is satisfied. For each selected model, the asynchronous state graph and the steady states (regular and singular) are given. Depending on the available biological knowledge, the user can

*   Reduce the domain of variation of some parameters,
*   Apply general constraints on parameters as, for example, the Snoussi's and observability[2] constraints,
*   Specify a set of steady states (regular and singular) and a set of functional circuits.

These direct constraints on parameters decrease significantly the number of models to generate and consequently increase the efficiency of the selection.

---

[1] Selection of Models for Biological Networks, see
http://smbionet.lami.univ-evry.fr

[2] Presented in the next section.

However, one can test directly the coherence of the regulatory graph (*i.e.* is there at least one suitable model ?), without enumeration of models by using a symbolic description of the set of all models.

In the next section, we shows how SMBioNet can be used for modelling the immunity control in bacteriophage lambda.

5.5 Immunity control in bacteriophage lambda

One of the most studied genetic regulatory networks is probably the one controlling immunity in temperate bacteriophage lambda, which is a temperate virus. As described in figure 5-12, after infection of a bacterial population,

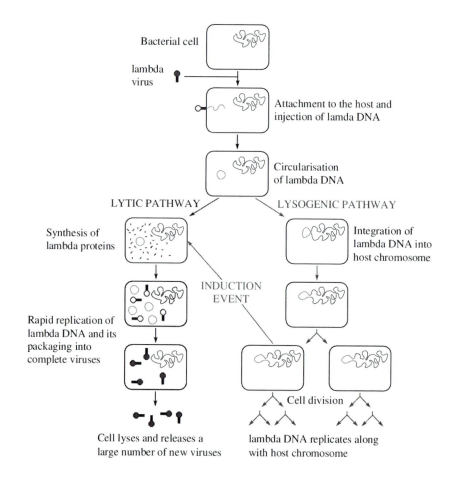

Figure 5-12. The life cycle of bacteriophage lambda.

many bacteria soon lyse and produce new phages but some survive and carry lambda genome in a dormant form. The first response is called *lytic* and the second *lysogenic*. In the lysogenic bacteria, viral DNA has integrated into the bacterial chromosome and will be faithfully transmitted to the bacterial progeny. In this condition, the viral gene cI, produces a repressor which blocks the expression of all the other genes of the phage, thus making the viral genome harmless for the bacterium. Moreover, cI makes lysogenic bacteria *immune* towards other infections. Lysogenization necessitates two events, integration of the viral DNA into the bacterial chromosome and development of immunity due to the expression of the repressor. The choice between the lytic and lysogenic pathways is very similar to cell differentiation, in the sense that a given virus, infecting apparently identical cells, can behave in two extremely different ways.

It is actually in the context of this biological system that Thomas started to develop his formalism. Although he proposed various models of the immunity control [26, 29, 30], we focus in this section on the model developed by Thieffry and Thomas in [26], which is denoted $\mathcal{M}(G)$ in the sequel. We will show that SMBioNet allows one to select, automatically and with very few biological knowledge, a set of models containing $\mathcal{M}(G)$ and satisfying the validation criteria given by Thieffry. All models of this set have to be considered since they have *a priori* the same prediction capacity than $\mathcal{M}(G)$.

5.5.1 Biological regulatory graph

The biological regulatory graph G summarizes the main regulations of the immunity control (Figure 5-13). Obviously it contains gene cI, but also three others (cro, cII, and N), which play a predominant role. Gene cI is activated by cII. Once on, gene cI remains on because its product activates its own synthesis, but at the same time, gene cI switches off the other lambda genes, including cII which had just switched it on. In addition gene cro exerts a negative control on cI, directly and indirectly, by repressing gene cII. Finally, gene N exerts a positive control on cII and is itself under negative control of cI and cII. According to the thresholds fixed by Thieffry, variables cI, cro, cII and N are 3-,4-,2- and 2-valued respectively, leading to 48 possible states. In the remainder, the state of the system is represented by the vector (cI,cro,cII,N).

5.5.2 Temporal properties

When the viral genome integrates a cell, all the viral proteins are initially absent. Thus (0,0,0,0) corresponds to the initial state of the system. The existence of both responses, lytic and lysogenic, implies that there exist two paths starting from the initial state leading respectively to the lytic state and to the immune one. The lytic state is known to be characterized by high concentration of cro and a low concentration of cI, cII and N whereas immune

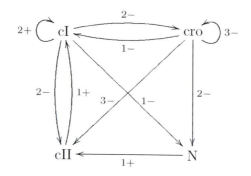

Figure 5-13. Biological regulatory graph $\mathcal{G}$ for immunity control.

state is characterized by high concentration of cI and low concentration of cro, cII and N. In [26], both states (0,2,0,0) and (0,3,0,0) correspond to the lytic state and (2,0,0,0) is the only state corresponding to the immunity. Without change of the environment, the choice between the lytic and the lysogenic pathways is irreversible, thus the lytic and immune states are steady. Then if the system reaches one state of the sets $A=\{(0,2,0,0),(0,3,0,0)\}$ or $B=\{(2,0,0,0)\}$, then it will never leave it. These sets of states are said steady sets.

Summing up, dynamics of models to consider have to contain paths from (0,0,0,0) to the steady sets of states A and B. These properties are translated into the CTL formula Φ as follow:

$$
\begin{aligned}
init &= ((cI = 0) \wedge (cro = 0) \wedge (cII = 0) \wedge (N = 0)) \\
lytic &= ((cI = 0) \wedge (cro \geq 2) \wedge (cII = 0) \wedge (N = 0)) \\
immune &= ((cI = 2) \wedge (cro = 0) \wedge (cII = 0) \wedge (N = 0)) \\
\Phi_A &= lytic \Rightarrow AG(lytic) \\
\Phi_B &= immune \Rightarrow AG(immune) \\
\Phi_r &= init \Rightarrow (\ EF(lytic) \wedge EF(immune)\ ) \\
\Phi &= \Phi_A \wedge \Phi_B \wedge \Phi_r
\end{aligned}
$$

The sub-formulae *init*, *lytic* and *immune* characterize the initial state, and the sets A and B. The steadiness of A and B is translated by Φ_A and Φ_B. The formula Φ_r expresses reachability of A and B from the initial state and Φ represents the temporal properties to use for the selection of models.

5.5.3 Selected models

There is near 7 thousands of millions of models associated to $\mathcal{G}$ leading to about 3 millions of different asynchronous state graphs. If we consider the

Snoussi's constraints (equation1) as Thieffry and Thomas did, it remains 151200 models. Moreover, we use the activity constraints [2]:

for each regulation $u \to v$ there is a set $\omega \subset G^-(v)$ such that $K_{v,\omega} \neq K_{v,\omega \cup \{u\}}$

which stands for the observability of any regulation. If $u \to v$ does not satisfy the constraints, the attractor of v does not depend on the level of u. It seems then quite obvious that any model should satisfy this property in order that all regulations play a role in the dynamics. Taking into account these constraints, SMBioNet selects among the 882 remaining models, 88 models satisfying the formula Φ. The model $\mathcal{M}(G)$ proposed by Thieffry and Thomas is one of them. Table 5-2 shows the possible values of parameters for the selected models. 17 parameters among 24 are fixed by formula Φ (in particular, all the parameters associated to N).

Table 5-2. Possible values of parameters for the selected models. Bold numbers correspond to the model $\mathcal{M}(G)$.

$K_{cI,\{\}}$	**= 0**	$K_{cII,\{\}}$	**= 0**
$K_{cI,\{cI\}}$	**= 1** or 2	$K_{cII,\{cI\}}$	**= 0**
$K_{cI,\{cro\}}$	= 0,1 or **2**	$K_{cII,\{cro\}}$	**= 0**
$K_{cI,\{cII\}}$	= 0,1 or **2**	$K_{cII,\{N\}}$	**= 0**
$K_{cI,\{cI,cro\}}$	**= 2**	$K_{cII,\{cI,cro\}}$	= 0 or **1**
$K_{cI,\{cI,cII\}}$	**= 1** or 2	$K_{cII,\{cI,N\}}$	= 0 or **1**
$K_{cI,\{cro\}}$	**= 2**	$K_{cII,\{cro,N\}}$	= 0 or **1**
$K_{cI,\{cI,cro,cII\}}$	**= 2**	$K_{cII,\{cI,cro,N\}}$	**= 1**
$K_{cro,\{\}}$	**= 0**	$K_{N,\{\}}$	**= 0**
$K_{cro,\{cI\}}$	**= 2**	$K_{N,\{cI\}}$	**= 0**
$K_{cro,\{cro\}}$	**= 0**	$K_{N,\{cro\}}$	**= 0**
$K_{cro,\{cI,cro\}}$	**= 2** or 3	$K_{N,\{cI,cro\}}$	**= 1**

5.5.4 Validation of models

Thieffry and Thomas exhibited one model whose coherence is analysed through the likelihood of some paths of the asynchronous state graph of $\mathcal{M}(G)$ and through the pertinence of predictions on the dynamics of some mutants. Our approach leads to select 88 models, which have to be evaluated with the same biological criteria of validation.

- Although 4 positive feedback circuits are present in the regulatory graph, the 88 selected models present only two steady states (regular or singular): (2,0,0,0) is always steady and the other one is either (0,2,0,0) or a singular state adjacent to (0,2,0,0) and (0,3,0,0). These steady states correspond to the lytic and immune states, and no other stable behaviour (phenotype) can be observed.

- Even if several pathways are possible from the initial state to immune state, all selected models present the most likely pathway in $M(G)$ from initial state to A (see Figure 5-13).

  Similarly the pattern of dynamics present in $M(G)$ allowing the system to evolve from initial state to lytic state, is also present in all selected models.

- Biological knowledge on mutants is available and can be used for validating models. The considered mutations correspond to the inactivation of different combinations of genes. Then simulations of the behaviour of these mutants can be performed and confronted to the biological knowledge. For example, the dynamics of the mutant $\lambda_{cI^-cro^-}$, where genes cI and cro are inactivated, is obtained from $M(G)$ by setting to 0 all parameters associated to cro or cI. Consequently, from an initial state where cI and cro are absent, they will never appear. The dynamics of this mutant is given in Figure 5-14.

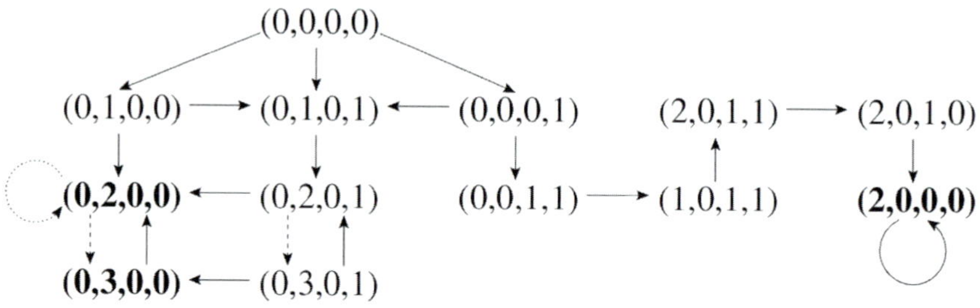

(0,0,0,0)

(0,1,0,0) ⟶ (0,1,0,1) ⟵ (0,0,0,1)   (2,0,1,1) ⟶ (2,0,1,0)

(0,2,0,0) ⟵ (0,2,0,1)   (0,0,1,1) ⟶ (1,0,1,1)   (2,0,0,0)

(0,3,0,0) ⟵ (0,3,0,1)

Figure 5-14. Likely paths from the initial state to the lytic and immune states (in bold). The dotted arrow is absent for the 44 models such that $K_{cro,\{cro,cI\}} = 3$, $M(G)$ included, whereas the dashed ones are absent for others.

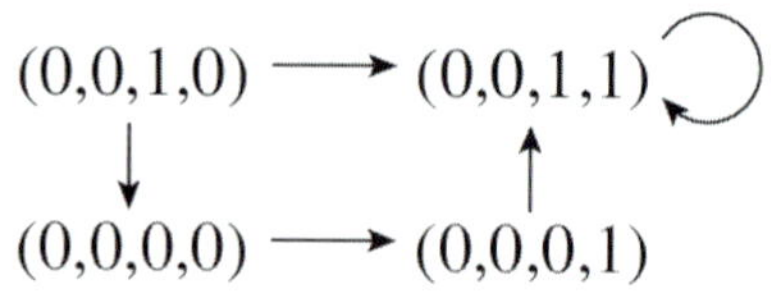

(0,0,1,0) ⟶ (0,0,1,1)

(0,0,0,0) ⟶ (0,0,0,1)

Figure 5-15. Dynamics of the mutant $\lambda_{cI^-cro^-}$ obtained from $M(G)$.

The dynamics of mutants obtained from $\mathcal{M}(\mathcal{G})$ are coherent from a biological point of view, since the remaining basins of attraction allow the prediction of the behaviour of mutants. For any selected model, results are the same and are given in table 5-3.

Among the 88 selected models, some differences can be highlighted. For example, 2 states are unreachable from the initial state in $\mathcal{M}(\mathcal{G})$ whereas for some selected models 15 states are unreachable. In such models, all states with cI $= 2$ and cro $= 3$ are not reachable, which is reasonable because high concentration of cI and cro is rarely observed. Moreover, such models do not contain the path $(0,0,0,0) \to (1,0,0,0) \to (2,0,0,0)$ present in the dynamics of $\mathcal{M}(\mathcal{G})$ and which is unlikely in view of the low expression of cI when cII is absent.

Table 5-3. Basins of attraction for a collection of mutants.

Mutants	Basins of attraction
λ_{cI^-}	A
λ_{cro^-}	B
λ_{cII^-}	A and B
λ_{N^-}	A and B
$\lambda_{cI^-cro^-}$	$\{(0,0,1,1)\}$
$\lambda_{cII^-N^-}$	A and B
$\lambda_{cro^-N^-}$	A
$\lambda_{cro^-cII^-}$	A
$\lambda_{cro^-cII^-N^-}$	A

In conclusion, these 88 selected models satisfy the same criteria of validation that $\mathcal{M}(\mathcal{G})$ and have also to be considered. These models have been selected using a formula Φ expressing the well known properties of the system. Thieffry and Thomas have exhibited their model with the circuit functionality and some hypothesis. We can notice that the used constraints for functionality are not necessary to reproduce the biological properties (expressed by Φ) because some of the models selected by Φ do not satisfy these functionality constraints. Moreover some parameters are valuated according to hypotheses ($K_{cII,\{cI,N\}} = 2$ for example) which have to be slacken since some models selected by Φ propose different values for these parameters.

5.6 Conclusion

We have defined a *formal* description of biological regulatory networks, which allows a computer aided manipulation of the semantics of the discrete modelling of Thomas, this manipulation being proved correct by construction. Our approach allows biology to take advantage of the whole corpus of formal methods from computer science. Model checking is a first powerful tool offered by the formalization of biological regulatory networks. In particular, temporal properties can be added into the specifications of the system, and the modelling task consists in exhibiting one or more generally all models that are coherent with the previous specifications expressing a part of the biological knowledge concerning the dynamics of the system. All potential models have to be checked against temporal formulae, and this task can be done automatically using model checking. This *brute force* approach permits one to exhibit exhaustively all suitable models, *i.e.* all models satisfying the temporal formulae. Information provided by a new experiment or a new theoretical point of view will refine the set of selected models.

The available temporal properties concern generally the homeostasis, the multi-stationarity, stable steady states and the accessibility of some stable steady states from a partially specified initial state. Unfortunately the stable steady states are some time singular and not formally represented in the asynchronous state graph of Thomas. Then the specifications cannot easily contain temporal properties concerning such singular states. This would necessitate to rewrite these temporal properties with only atomic propositions of regular states, and this task is generally difficult.

De Jong et al. [7] introduced the singular states into their qualitative dynamics. Their qualitative modelling of genetic regulatory networks is also based on piecewise-linear differential equations. Authors propose a mathematically well founded method to deal with singular states using differential inclusions [9, 11]. Our approach consisting in adding temporal properties into the specifications for determining the suitable parameter values, would allows in this context to treat regular states as well as singular states.

More generally the formal methods can be applied in the field of biological regulatory networks and systems biology in order to explicit some behaviours or to take into account biological knowledge which have been ignored for the moment. The cooperation between biology and formal methods from computer science opens a large horizon of research perspectives.

- The introduction of transitions in the regulatory graph could help to specify how the different regulators cooperate for inducing or repressing their common target [1]. One can also separate inhibitors from activators [2] to increase the expressivity of the approach, or take into account time delays [31] between the beginning of the activation order and the synthesis of the product and conversely for the turn-off delays.

- Automatic generation of experiment schema from suitable models. In order to reduce again the set of suitable models, we would like to propose the biologist to perform an determining experiment. The result is then confronted to each model and only those, which are coherent with the experiment, have to be kept. An experiment often consists to put the system in a particular state (partially specified) and to observe after a while if one or several gene products are present or not. This implies to extract the specificities of the biological application domain in order to define patterns of formulae expressing feasible experiments.
- The modelling of a regulatory network concerns generally only a small part of the global regulatory network of the cell. It becomes crucial to prove that the dynamical properties of this sub-network are preserved when it is embedded into the global network. This is correlated to the treatment of knock-out mutants, identification of functional patterns [21] as well as the structure of huge regulatory networks.

To achieve such development several directions have to be considered. High-level Petri nets are graphical oriented languages for design, specification, simulation and verification of systems. They are in particular well-suited for systems in which communication, synchronization and resource sharing are important. Clearly, biological systems present these characteristics, and modelling by such nets would allows us to take advantage of all results and tools in the field of high-level Petri nets.

Hybrid automata can take into account the continuous aspects of a regulatory network: it is a mathematical model for hybrid systems, which combines, in a single formalism, automaton transitions for capturing discrete changes with differential equations for capturing continuous changes. Symbolic model checkers, as HyTech [12], have been developed for the subclass of linear hybrid automata. It becomes possible to perform parametric analysis, *i.e.* to determine the values of parameters for which a linear hybrid automaton satisfies a temporal-logic requirement.

These research perspectives aim to link modelling and experiments together, by furnishing to biologists model structuring methods and model validation tools from current researches in theoretical computer science. The resulting formal models are not only *a posteriori* explanations of biological results, they are guides for biological experiments whose success will be *in fine* the discriminating criterion.

5.7 Acknowledgements

The authors thank *genopole*®-research in Evry (H. Pollard and P. Tambourin) for constant supports. We gratefully acknowledge the members of the *genopole*® working groups *observability* and G^3 for stimulating interactions.

5.8 References

[1]   V. Bassano and G. Bernot. Marked regulatory graphs: a formal framework to simulate biological regulatory networks with simple automata. In *14'th International Workshop on Rapid System Prototyping*, pages 93-99, San Diego, 2003.

[2]   G. Bernot, F. Cassez, J.-P. Comet, F. Delaplace, C. Müller, O. Roux, and O.H. Roux. Semantics of biological regulatory networks. In *Proceedings of the Workshop on Concurrent Models in Molecular Biology (BioConcur'2003)*, 2003.

[3]   G. Bernot, J.-P. Comet, A. Richard, and J. Guespin. A fruitful application of formal methods to biological regulatory networks: Extending Thomas' asynchronous logical approach with temporal logic. *J. Theor. Biol.*, 229(3):339-347, 2004.

[4]   A. Cimatti, E. Clarke, F. Giunchiglia, and M. Roveri. NuSMV: a reimplementation of SMV. In *Proceeding of the International Workshop on Software Tools for Technology Transfer (STTT-98)*, BRICS Notes Series, NS-98-4, pages 25-31, 1998.

[5]   O. Cinquin and J. Demongeot. Positive and negative feedback: striking a balance between necessary antagonists. *J. Theor. Biol.*, 216(2):229-241, 2002.

[6]   O. Cinquin and J. Demongeot. Roles of positive and negative feedback in biological systems. *C.R.Biol.*, 325(11):1085-1095, 2002.

[7]   H. de Jong, J.-L. Gouzé, C. Hernandez, M. Page, S. Tewfik, and J. Geiselmann. Qualitative simulation of genetic regulatory networks using piecewise-linear models. *Bull. Math. Biol.*, 66(2):301-340, 2004.

[8]   E.A. Emerson. *Handbook of theoretical computer science, Volume B : formal models and semantics*, chapter Temporal and modal logic, pages 995-1072. MIT Press, 1990.

[9]   A.F. Filippov. *Differential equations with discontinuous right-hand sides*. Kluwer Academic Publishers, 1988.

[10]   L. Glass and S.A. Kauffman. The logical analysis of continuous non linear biochemical control networks. *J. Theor. Biol.*, 39(1):103-129, 1973.

[11]   J.-L. Gouzé and S. Tewfik. A class of piecewise linear differential equations arising in biological models. *Dynamical Syst.*, 17:299-316, 2003.

[12]   T.A. Henzinger, P.-H. Ho, and H. Wong-Toi. HyTech: A model checker for hybrid systems. *Software Tools for Technology Transfer*, 1:110-122, 1997.

[13]   M. Huth and M. Ryan. *Logic in Computer Science: Modelling and reasoning about systems*. Cambridge University Press, 2000.

[14]   B.J. Kuipers. *Qualitative reasoning: modeling and simulation with incomplete knowledge*. MIT Press, 1994.

[15]   K. McMillan. *Symbolic Model Checking*. Kluwer Academic Publishers, 1993.

[16] L. Mendoza, D. Thieffry, and E.R. Alvarez-Buylla. Genetic control of flower morphogenesis in arabidopsis thaliana: a logical analysis. *Bioinformatics*, 15(7-8):593-606, 1999.

[17] S. Pérès and J.-P. Comet. Contribution of computational tree logic to biological regulatory networks: example from pseudomonas aeruginosa. In *International workshop on Computational Methods in Systems Biology*, volume 2602 of *LNCS*, pages 47-56, February 24-26, 2003.

[18] E. Plathe, T. Mestl, and S.W. Omholt. Feedback loops, stability and multistationarity in dynamical systems. *J. Biol. Syst.*, 3:569-577, 1995.

[19] L. Sánchez and D. Thieffry. A logical analysis of the drosophila gap-gene system. *J. Theor. Biol.*, 211(2):115-141, 2001.

[20] L. Sánchez, J. van Helden, and D. Thieffry. Establishment of the dorso-ventral pattern during embryonic development of drosophila melanogaster: a logical analysis. *J. Theor. Biol.*, 189(4):377-389, 1997.

[21] S.S. Shen-Orr, R. Milo, S. Mangan, and U. Alon. Network motifs in the transcriptional regulation network of Escherichia coli. *Nat. Genet.*, 31(1):64-68, 2002.

[22] E.H. Snoussi. Qualitative dynamics of a piecewise-linear differential equations : a discrete mapping approach. *Dynamics and stability of Systems*, 4:189-207, 1989.

[23] E.H. Snoussi. Necessary conditions for multistationarity and stable periodicity. *J. Biol. Syst.*, 6:3-9, 1998.

[24] E.H. Snoussi and R. Thomas. Logical identification of all steady states : the concept of feedback loop characteristic states. *Bull. Math. Biol.*, 55(5):973-991, 1993.

[25] C. Soulé. Graphical requirements for multistationarity. *ComPlexUs*, 1:123-133, 2003.

[26] D. Thieffry and R. Thomas. Dynamical behaviour of biological regulatory networks - II. Immunity control in bacteriophage lambda. *Bull. Math. Biol.*, 57(2):277-297, 1995.

[27] R. Thomas. Logical analysis of systems comprising feedback loops. *J. Theor. Biol.*, 73(4):631-656, 1978.

[28] R. Thomas. On the relation between the logical structure of systems and their ability to generate multiple steady states or sustained oscillations. *Springer Series in Synergies 9*, pages 180-193, 1980.

[29] R. Thomas and R. d'Ari. *Biological Feedback*. CRC Press, 1990.

[30] R. Thomas, A.M. Gathoye, and L. Lambert. A complex control circuit. Regulation of immunity in temperate bacteriophages. *Eur. J. Biochem.*, 71(1):211-227, 1976.

[31] R. Thomas and M. Kaufman. Multistationarity, the basis of cell differentiation and memory. II. Logical analysis of regulatory networks in terms of feedback circuits. *Chaos*, 11:180-195, 2001.

[32] R. Thomas, D. Thieffry, and M. Kaufman. Dynamical behaviour of biological regulatory networks - I. Biological role of feedback loops an practical use of the concept of the loop-characteristic state. *Bull. Math. Biol.*, 57(2):247-276, 1995.

[33] E.O. Voit. *Computational Analysis of biochemical systems: a practical guide for biochemists and molecular biologists*. Cambridge University Press, 2000.

6 Formal Methods for Specifying and Analyzing Complex Software Systems

Author
Xudong He[1], Huiqun Yu[2], and Yi Deng[1]

[1]School of Computer Science
Florida International University, U.S.A.

[2]Department of Computer Science and Engineering
East China University of Science and Technology, China

Summary
Software has been a major enabling technology for advancing modern society, and is now an indispensable part of daily life. Because of the increased complexity of these software systems, and their critical societal role, more effective software development and analysis technologies are needed. How to develop and ensure the dependability of these complex software systems is a grand challenge.

It is well-known that a highly dependable complex software system cannot be developed without a rigorous development process and a precise specification and design documentation. Formal methods are one of the most promising technologies for precisely specifying, modeling, and analyzing complex software systems. Although past research experience and practice in computer science have convincingly shown that it is not possible to formally verify program behavior and properties at the program source code level due to its extreme huge size and complexity, recently advances in applying formal methods during software specification and design, especially at software architecture level, have demonstrated significant benefits of using formal methods.

In this chapter, we will review several well-known formal methods for software system specification and analysis. We will present recent advances of using these formal methods for specifying, modeling and analyzing software architectural design.

Hossam A. Gabbar (ed.), Modern Formal Methods and Applications, 123–150.
© 2006 *Springer. Printed in the Netherlands.*

6.1 Introduction

It is wildly agreed that the main obstacle to "help computers help us more" and relegate to these helpful partners even more complex and sensitive tasks is not inadequate speed and unsatisfactory raw computing power in the existing machines, but our limited ability to design and implement complex systems with sufficiently high degree of confidence in their correctness under all circumstances [CGP99]. This problem of design validation – ensuring the correctness of the design at the earliest stage possible – is the major challenge in any responsible system development process, and the activities intended for its solution occupy an ever increasing portion of the development cycle cost and time budgets.

Two major approaches to analyze the system quality are testing and verification. Traditional and widely used quality assurance techniques based on software testing are inadequate to ensure the reliability of complex systems. In addition to the inherent limitation of testing from being able to guarantee system properties, many of today's software systems are designed to adapt in a wide range of environments and evolve over time. Because of this, the range of possible testing scenarios at code level becomes extremely large and potentially uncontrollable.

Formal methods [Har87, Hoa85, Mil89, MP92, Mur89] for software specification and verification have been viewed as a promising way to address the problems associated with testing. These methods are precise and rigorous and can prevent and detect system defects introduced at the early stages of development, which are often more costly to fix and have more severe consequences. Despite of tremendous advances, however, widely spread application of formal methods in practical system development still remains to be seen [CGR95]. A major cause for the problem is that results on formal methods are to large extent fragmented. Formal techniques are viewed as difficult and expensive to use because their application is ad hoc, and they are too fine grained to deal with the complexity in practical-sized development. Thus it is necessary to precisely define, measure, and analyze software dependability at a level higher than source code. Recent research has shown that it is especially important to explore technologies how to handle dependability attributes at the software architecture level for the following reasons:

- A software architecture description presents the highest-level design abstraction of a system. As a result it is relative simple compared to a detailed system design. Thus it is more likely to develop an effective methodology to study dependability attributes.
- As the highest-level design abstraction, a software architecture description precedes and logically and structurally influences other system development products. Thus an error in a software architecture has a much

larger impact than an error introduced at a later development stage. Prevention and detection of errors at software architectural level are thus extremely important. Hence, it is necessary to study and measure dependability attributes before the actual software systems are developed and deployed.

Many studies, especially those done at the Software Engineering Institute at Carnegie Mellon University, have shown that a software architecture reveals, influences, or even dictates many system dependability features such as reliability, performance, security, and faulty-tolerance. Therefore, the dependability attributes measured at software architecture level can serve as the basis to predict and validate the dependability attributes of the developed and deployed systems.

In this chapter, we will review several well-known formal methods for complex software system specification and analysis. We will illustrate these methods and their applications in the Software Architecture Model (SAM) [WHD99, HD02], which is a general software architecture model for developing and analyzing software architecture specifications.

6.2 Formal Specification Techniques

6.2.1 Visualizing the Structures of Software Architectures

Specification is the process of describing a system and its desired properties. Formal specification uses a language that is usually composed of three primary components: (1) a *syntax* that defines the specific notation with which the specification is represented, (2) a *semantics* that helps to define a "universe of objects" [Win90] that will be used to describe the system, and (3) a set of *relations* that define the rules that indicate which objects properly satisfy the specification.

In SAM, a software architecture is visualized by a hierarchical set of boxes with ports connected by directed arcs. These boxes are called *compositions*. Each composition may contain other compositions. The bottom-level compositions are either *components* or *connectors*. Various constraints can be specified. This hierarchical model supports compositionality in both software architecture design and analysis, and thus facilitates scalability. Figure 6-1 shows a graphical view of an SAM software architecture, in which connectors are not emphasized and are only represented by thick arrows. Each component or connector is defined using a Petri net. Thus the internal logical structure of a component or connector is also visualized through the Petri net structure.

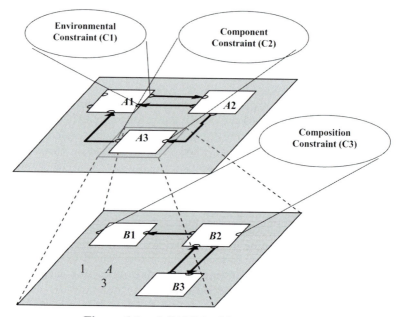

Figure 6-1. –A SAM Architecture Model

Textually, an SAM software architecture is defined by a set of compositions C = $\{C_1, C_2, ...,C_k\}$ (each composition corresponds to a design level or the concept of sub-architecture) and a hierarchical mapping h relating compositions. Each composition $C_i = \{Cm_i, Cn_i, Cs_i\}$ consists of a set Cm_i of components, a set Cn_i of connectors, and a set Cs_i of composition constraints. An element $C_{ij} = (S_{ij}, B_{ij})$, (either a component or a connector) in a composition C_i has a property specification S_{ij} (a temporal logic formula) and a behavior model B_{ij} (a Petri net). Each composition constraint in Cs_i is also defined by a temporal logic formula. The interface of a behavior model B_{ij} consists of a set of places (called ports) that is the intersection among relevant components and connectors. Each property specification S_{ij} only uses the ports as its atomic propositions / predicates that are true in a given marking if they contain appropriate tokens. A composition constraint is defined as a property specification, however it often contains ports belonging to multiple components and / or connectors. A component C_{ij} can be refined into a lower-level composition C_l, which is defined by $h(C_{ij}) = C_l$.

6.2.2   Modeling the Behaviors of Software Architectures

In SAM, the behavior of a component or a connector is explicitly defined using a Petri net. The behavior of an overall software architecture is implicitly derived by composing all the bottom-level behavior models of components and connectors. SAM provides both the modeling power and flexibility through the choice of different Petri net models. We have used several Petri net models including time Petri nets [WHD99], condition event nets, and predicate transition nets [HD00, HD02] in our previous work. The selection of a particular Petri net model is based on the application under consideration. A simple Petri net model such as condition event nets is adequate when we only need to deal with simple control flows and data-independent constraints; while a more powerful Petri net model such as predicate transition nets is needed to handle both control and data. To study performance related constraints, a more specialized Petri net model such as stochastic Petri nets is more appropriate and convenient. In the following sections, we give a brief definition of predicate transition nets (PrT nets) using the conventions in [He96].

6.2.3   The Syntax and Static Semantics of PrT Nets

A PrT net is a tuple $(N, Spec, ins)$ where
(1) $N = (P, T, F)$ is the net structure, in which
     (i) P and T are non-empty finite sets satisfying $P \cap T = \emptyset$ (P and T are the sets of places and transitions of N respectively),
     (ii) $F \subseteq (P \times T) \cup (T \times P)$ is a flow relation (the arcs of N);
(2) $Spec = (S, OP, Eq)$ is the underlying specification, and consists of a signature $\mathbf{S} = (S, OP)$ and a set Eq of S-equations. Signature $\mathbf{S} = (S, OP)$ includes a set of sorts S and a family $OP = (OP_{s_1,\ldots,s_n, s})$ of sorted operations for $s_1, \ldots, s_n, s \in S$. For each $s \in S$, we use CON_S to denote $OP_{,s}$ (the 0-ary operation of sort s), i.e. the set of constant symbols of sort s. The S-equations in Eq define the meanings and properties of operations in OP. We often simply use familiar operations and their properties without explicitly listing the relevant equations. $Spec$ is a meta-language to define the tokens, labels, and constraints of a PrT net. Tokens of a PrT net are ground terms of the signature S, written $MCON_S$. The set of labels is denoted using $Label_S(X)$ (X is the set of sorted variables disjoint with OP). Each label can be a multiple set expression of the form $\{k_1 x_1, \ldots, k_n x_n\}$. Constraints of a PrT net are a subset of first order logic formulas (where the domains of quantifiers are finite and any free variable in a constraint appears in the label of some connecting arc of the transition), and thus are essentially propositional logic formulas. The subset of first order logical formulas contains the S-terms of sort *bool* over X, denoted as $Term_{OP,bool}(X)$.

(3) $ins = (\varphi, L, R, M_0)$ is a net inscription that associates a net element in N with its denotation in $Spec$:
     (i) $\varphi: P \rightarrow \wp(S)$ is the data definition of N and associates each place p in P with a subset of sorts in S.

(ii) $L: F \rightarrow Label_S(X)$ is a sort-respecting labeling of PrT net. We use the following abbreviation in the following definitions:

$$\overline{L}(x,y) \qquad \begin{array}{ll} L(x,y) & \text{iff } (x,y) \quad F \\ & \text{otherwise} \end{array}$$

(iii) $R: T \rightarrow Term_{OP,bool}(X)$ is a well-defined constraining mapping, which associates each transition t in T with a first order logic formula defined in the underlying algebraic specification. Furthermore, the constraint of a transition defines the meaning of the transition.

(iii) $M_0: P \rightarrow MCON_S$ is a sort-respecting initial marking. The initial marking assigns a multi-set of tokens to each place p in P.

6.2.4 Dynamic Semantics of PrT Nets

(1) Markings of a PrT net N are mappings $M: P \rightarrow MCON_S$;

(2) An occurrence mode of N is a substitution $\alpha = \{x_1 \leftarrow c_1, \ldots, x_n \leftarrow c_n\}$, which instantiates typed label variables. We use $e:\alpha$ to denote the result of instantiating an expression e with α, in which e can be either a label expression or a constraint;

(3) Given a marking M, a transition $t \in T$, and an occurrence mode α, t is α_enabled at M iff the following predicate is true: $\forall p: p \in P.(\overline{L}(p,t):\alpha) \subseteq M(p)) \wedge R(t):\alpha$;

(4) If t is α_enabled at M, t may fire in occurrence mode α. The firing of t with α returns the marking M' defined by $M'(p) = M(p) - \overline{L}(p,t):\alpha \cup \overline{L}(t,p):\alpha$ for $p \in P$. We use $M[t/\alpha>M'$ to denote the firing of t with occurrence α under marking M. As in traditional Petri nets, two enabled transitions may fire at the same time as long as they are not in conflict;

(5) For a marking M, the set $[M>$ of markings reachable from M is the smallest set of markings such that $M \in [M>$ and if $M' \in [M>$ and $M'[t/\alpha>M''$ then $M'' \in [M>$, for some $t \in T$ and occurrence mode α (note: concurrent transition firings do not produce additional new reachable markings);

(6) An execution sequence $M_0T_0M_1T_1\ldots$ of N is either finite when the last marking is terminal (no more enabled transition in the last marking) or infinite, in which each T_i is an execution step consisting of a set of non-conflict firing transitions;

(7) The behavior of N, denoted by *Comp(N)*, is the set of all execution sequences starting from the initial marking.

The Dining Philosophers problem is a classic multi-process synchronization problem introduced by Dijkstra. The problem consists of k philosophers sitting at a round table who do nothing but think and eat. Between each philosopher, there is a single chopstick. In order to eat, a philosopher must have both chopsticks. A problem can arise if each philosopher grabs the chopstick on the right, then waits for the stick on the left. In this case a deadlock has occurred.

The challenge in the Dining Philosophers problem is to design a protocol so that the philosophers do not deadlock (i.e. the entire set of philosophers does not stop and wait indefinitely), and so that no philosopher starves (i.e. every philosopher eventually gets his/her hands on a pair of chopsticks). The following is an example of the PrT net model of the Dining Philosophers problem.

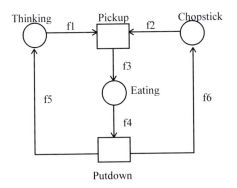

Figure 6-2. – A PrT Net Model of the Dining Philosophers Problem

There are three places (*Thinking, Chopstick* and *Eating*) and two transitions (*Pickup* and *Putdown*) in the PrT net. In the underlying specification *Spec* = (*S, OP, Eq*), *S* includes elementary sorts such as Integer and Boolean, and also sorts PHIL and CHOP derived from Integer. *S* also includes structured sorts such as set and tuple obtained from the Cartesian product of the elementary sorts; *OP* includes standard arithmetic and relational operations on Integer, logical connectives on Boolean, set operations, and selection operation on tuples; and *Eq* includes known properties of the above operators.

The net inscription (φ, L, R, M_0) is as follows:
- Sorts of predicates:
  $\varphi(Thinking) = \wp(\text{PHIL})$, $\varphi(Eating) = \wp(\text{PHIL} \times \text{CHOP} \times \text{CHOP})$,
  $\varphi(Chopstick) = \wp(\text{CHOP})$,
  where $\wp$ denotes power set.
- Arc definitions:
  $L(f1) = \{ph\}$, $L(f2) = \{ch1,ch2\}$, $L(f3) = \{<ph,ch1,ch2>\}$,
  $L(f4) = \{<ph,ch1,ch2>\}$, $L(f5) = \{ph\}$, $L(f6) = \{ch1,ch2\}$.
- Constraints of transitions:
  $R(Pickup) = (ph = ch1) \wedge (ch2 = ph \oplus 1)$, $R(Putdown) = \text{true}$.
- The initial marking m_0 is defined as follows:
  $M_0(Thinking) = \{1, 2, ..., k\}$, $M_0(Eating) = \{ \}$, $M_0(Chopstick) = \{1, 2, ..., k\}$.

The above specification allows concurrent executions such as multiple non-conflicting (non-neighboring) philosophers picking up chopsticks simultaneously, and some philosophers picking up chopsticks while others putting down chopsticks. The constraints associated with transitions Pickup and Putdown also ensure that a philosopher can only use two designated chopsticks defined by the implicit adjacent relationships. Table 6-1 gives the details of a possible run of five dining philosophers PrT net.

Table 6-1. A Possible Run of Five Dining Philosophers Problem

Markings m_i			Transitions n_i	
Thinking	Eating	Chopstick	Fired Transition	Token(s) consumed
{1,2,3,4,5}	{ }	{1,2,3,4,5}	Pickup	ph=1, ch1=1, ch2=2
{2,3,4,5}	{<1,1,2>}	{3,4,5}	Putdown	<ph,ch1,ch2>=<1,1,2>
{1,2,3,4,5}	{ }	{1,2,3,4,5}	Pickup	ph=2, ch1=2, ch2=3
{1,3,4,5}	{<2,2,3>}	{1,4,5}	Pickup	ph=4, ch1=4, ch2=5
{1, 3, 5}	{<2,2,3>, <4,4,5>}	{1}	Putdown	<ph,ch1,ch2>=<2,2,3>
{1, 2, 3, 5}	{<4, 4, 5>}	{1,2,3}	Putdown	<ph,ch1,ch2>=<4,4,5>
{1,2,3,4,5}	{ }	{1,2,3,4,5}	Pickup	ph=5, ch1=5, ch2=1
{1,2,3,4}	{<5,5,1>}	{2,3,4}	Pickup	ph=3, ch1=3, ch2=4
{1,2,4}	{<5,5,1>, <3,3,4>}	{2}	Putdown	<ph,ch1,ch2>=<3,3,4>
{1,2,3,4}	{<5,5,1>}	{2,3,4}	Putdown	<ph,ch1,ch2>=<5,5,1>
{1,2,3,4,5}	{ }	{1,2,3,4,5}	...	...

6.2.5 Specifying SAM Architecture Properties

In SAM, software architecture properties are specified using a temporal logic. Depending on the given Petri net models, different temporal logics are used. In this section, we provide the essential concepts of a generic first order linear time temporal logic to specify the properties of components and connectors. We follow the approach in [Lam94] to define vocabulary and models of our temporal logic in terms of PrT nets without giving a specific temporal logic.

6.2.5.1   Values, State Variables, and States

The set of values is the multi-set of tokens *MCONS* defined by the *Spec* of a given PrT net *N*. Multi-sets can be viewed as partial functions. For example, multi-set $\{3a, 2b\}$ can be represented as $\{a \mapsto 3, b \mapsto 2\}$.

The set of state variables is the set P of places of N, which change their meanings during the executions of N. The arity of a place p is determined by its sort $\varphi(p)$ in the net inscription.

The set of states **St** is the set of all reachable markings $[M_0>$ of N. A marking is a mapping from the set of state variables into the set of values. We use $M[|x|]$ to denote the value of x under state (marking) M.

Since state variables take partial functions as values, they are flexible function symbols. We can access a particular component value of a state variable. However there is a problem associated with partial functions, i.e. many values are undefined. The above problem can easily be solved by extending state variables into total functions in the following way: for any n-ary state variable p, any tuple $c \in MCON_s^n$ and any state M, if $p(c)$ is undefined under M, then let $M[|p(c)|] = 0$. The above extension is consistent with the semantics of PrT nets. Furthermore we can consider the meaning $[|p(c)|]$ of the function application $p(c)$ as a mapping from states to **Nat** using a postfix notation for function application $M[|p(c)|]$.

6.2.5.2   Rigid Variables, Rigid Function and Predicate Symbols

Rigid variables are individual variables that do not change their meanings during the executions of N. All rigid variables occurring in our temporal logic formulas are bound (quantified), and they are the only variables that can be quantified. Rigid variables are variables appearing in the label expressions and constraints of N. Rigid function and predicate symbols do not change their meanings during the executions of N. The set of rigid function and predicate symbols is defined in the *Spec* of N.

6.2.5.3   State Functions, Predicates, and Transitions

A *state function* is an expression built from values, state variables, rigid function and predicate symbols. For example $[|p(c) + 1|]$ is a state function where c and 1 are values, p is a state variable, $+$ is a rigid function symbol. Since the meanings of rigid symbols are not affected by any state, thus for any given state M, $M[|p(c) + 1|] = M[|p(c)|] + 1$.

A *predicate* is a boolean-valued state function. A predicate p is said to be satisfied by a state M iff $M[|p|]$ is true.

A *transition* is a particular kind of predicates that contain primed state variables, e.g. $[|p'(c) = p(c) + 1|]$. A transition relates two states (an old state and a new state), where the unprimed state variables refer to the old state and the primed state variables refer to the new state. Therefore the meaning of a transition is a relation between states. The term transition used here is a temporal logic entity. Although it reflects the nature of a transition in a PrT net N, it is not a transition in N. For example, given a pair of states M and M': $M[|p'(c) = p(c) + 1|]M'$ is defined by $M'[|p'(c)|] = M[|p(c)|] + 1$. Given a transition t, a pair of states M and M' is called a "transition step" iff $M[|t|]M'$ equals true. We can easily generalize any predicate p without primed state variables into a relation between states by replacing all unprimed state variables with their primed versions such that $M[|p'|]M'$ equals $M'[|p|]$ for any states M and M'.

6.2.5.4 Temporal Formulas

Temporal formulas are built from elementary formulas (predicates and transitions) using logical connectives $\neg$ and $\wedge$ (and derived logical connectives $\vee$, $\Rightarrow$, and $\Leftrightarrow$), universal quantifier $\forall$ and derived existential quantifier $\exists$, and temporal operators always $\Box$, sometimes $\Diamond$, and until U.

The semantics of temporal logic is defined on behaviors (infinite sequences of states). The behaviors are obtained from the execution sequences of PrT nets where the last marking of a finite execution sequence is repeated infinitely many times at the end of the execution sequence. For example, for an execution sequence $M_0,...,M_n$, the following behavior $\sigma = <<M_0,...,M_n,M_n,... >>$ is obtained. We denote the set of all possible behaviors obtained from a given PrT net as $\mathbf{St}^\infty$.

Let u and v be two arbitrary temporal formulas, p be an n-ary predicate, t be a transition, x, $x_1,...,x_n$ be rigid variables, $\sigma = <<M_0, M_1, ... >>$ be a behavior, and $\sigma^k = <<M_k, M_{k+1}, ... >>$ be a k step shifted behavior sequence; we define the semantics of temporal formulas recursively as follows:

$(1)\ \sigma\,[|p(x_1,...,x_n)|]\quad \equiv \qquad M_0[|\,p(x_1,...,x_n)|]$

$(2)\ \sigma\,[|t|]\qquad \equiv \qquad M_0[|\,t|]M_1$

$(3)\ \sigma\,[|\neg u|]\qquad \equiv \qquad \neg\,\sigma\,[|u|]$

$(4)\ \sigma\,[|u \wedge v|]\qquad \equiv \qquad \sigma\,[|u|] \wedge \sigma\,[|\,v\,|]$

$(5)\ \sigma\,[|\forall x.\,u|]\qquad \equiv \qquad \forall x.\sigma\,[|u|]$

$(6)\ \sigma\,[|\Box\,u|]\qquad \equiv \qquad \forall n \in \mathbf{Nat}.\ \sigma^n\,[|u|]$

$(7)\ \sigma\,[|uUv|]\qquad \equiv \qquad \exists\,k.\,\sigma^k\,[|v|] \wedge \forall\,0 \le n \le k.\,\sigma^n\,[|u|]$

A temporal formula u is said to be *satisfiable*, denoted as $\sigma \models u$, iff there is an execution σ such that $\sigma [|u|]$ is true, i.e. $\sigma \models u \Leftrightarrow \exists \sigma \in \mathbf{St}^\infty. \sigma [|u|]$. u is *valid* with regard to N, denoted as $N \models u$, iff it is satisfied by all possible behaviors $\mathbf{St}^\infty$ from N: $N \models u \Leftrightarrow \forall \sigma \in \mathbf{St}^\infty. \sigma [|u|]$.

2.3.5 Defining System Properties in Temporal Logic

Specifying architecture properties in SAM becomes defining PrT net properties using temporal logic. Canonical forms for a variety of system properties such as safety, guarantee, obligation, response, persistence, and reactivity are given in [MP92]. For example, the following temporal logic formulas specify a safety property and a liveness property of the PrT net in Fig.2 respectively:

- Mutual exclusion:

$$\forall ph \in \{1, ..., k\} \square \neg (< ph, \_, \_ > \in Eating \wedge < ph \oplus 1, \_, \_ > \in Eating),$$

  which defines that no adjacent philosophers can eat at the same time.
- Starvation freedom: $\forall ph \in \{1, ..., k\} \lozenge (< ph, \_, \_ > \in Eating)$,

  which states that every philosopher will eventually get a chance to eat.

6.3 Formal Methods for Designing Software Architectures

There are two distinct levels of software architecture specification development in SAM: element level and composition level. The element level specification deals with the specification of a single component or connector, and the composition level specification concerns how to combine (horizontal) specifications at the same abstraction level together and how to relate (vertical) specifications at different abstraction levels.

6.3.1 Developing Element Level Specifications

In SAM, each element (either a component or a connector) is specified by a tuple $<S, B>$. S is a property specification, written in temporal logic, that specifies the required properties of the element, and B is a behavior model, defined by a PrT net, that defines the behavior of the element. S and B can be viewed as the specification and the implementation respectively as in many other software architecture models such as Wright [AG97]. Therefore to develop the specification of an element is essentially to write S and B.

Although many existing techniques for writing temporal logic specifications [MP92, Lam94] and for developing Petri nets [Rei92, HY92, Jen92] may be directly used here. There are several unique features about <S, B>. First, S and B are related and constrain each other. Thus we have to develop either S or B with respect to a possibly existing B or S. Depending on our understanding of a given system; we can either develop S or B first. Second, the predicate symbols used in S are exterior (either input or out) ports of B. Third, S should in general be weaker than B, i.e. B may satisfy more properties than S. Thus the view of implementation as implication is valid here. With the above unique features in mind, we offer the following heuristics for developing S and B:

Heuristic 1 - How to Write *S*

To define an element constraint, we can either directly formulate the given user requirements or carry out a cause effect analysis by viewing input ports as causes and output ports as effects. Canonical forms [MP92] for a variety of properties such as Safety, Guarantee, Obligation, Response, Persistence, and Reactivity are used as guidelines to define property specifications.

A simple example of applying Heuristic 1 is as follows. Let us consider a simple automated library system that supports typical transaction types such as checkout and return a book. A transaction is initiated with a user request that contains user identification, a book title, and a transaction type (checkout / return). The transaction is processed by updating the user record and the book record, and is finished by sending the user a message – either successful or a failure reason. One desirable property of an automated library system is that each request must be proposed. This property is a type of response property [MP92], and thus can be defined as

$$\forall(req).(\Box(\text{Request}(req) \Rightarrow \Diamond\text{Response}(msg))),$$

where *req* and *msg* stand for a request and message (Success or Failure) respectively, and Request and Response are predicate symbols, and must correspond to an input port and an output port respectively.

Heuristic 2 - How to Develop *B*
We follow the general procedure proposed in [HY92] to develop B.
Step 1 - use all the input and output ports as places of B;
Step 2 - identify a list of events directly from the given user requirements or through Use Case analysis [BRJ99];
Step 3 - represent each event with a simple PrT net;
Step 4 - merge all the PrT nets together through shared places to obtain B;
Step 5 - apply the transformation techniques [HL91] to make B more structured and / or meaningful.

Again, we use the above simple library system as an example. We only provide a partial behavior model without the complete net inscription to illustrate the application of Heuristic 2. A more complete example of a PrT net specification of a library system can be found in [HY92]. Since we developed a property specification first in this case and we identified an input port Request and an output port Response, we use them as places in the behavior model *B* according to Step 1. We can easily identify two distinct types of events: checkout and return. According to Step 3, we come up with the following two PrT nets Figure 3 (a) and (b), each of which models an event type. Figure 3 (c) is obtained by merging shared places according Step 4, and Figure 3 (d) is obtained by restructuring Figure 6-3 (c) through combining Checkout and Return into a generic transaction type.

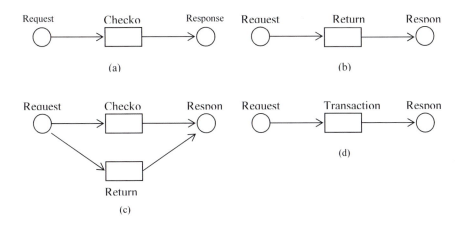

Figure 6-3. (a) A PrT model of Checkout; (b) a PrT model of Return;
(c) a connected PrT model; (d) A PrT model of Checkout

6.3.2 Developing Composition Level Specifications

SAM supports both top-down and bottom-up system development approaches. The top-down approach is used to develop a software architecture specification by decomposing a system specification into specifications of components and connectors and by refining a higher level component into a set of related sub-components and connectors at a lower level. The bottom-up approach is used to develop a software architecture specification by composing existing specifications of components and connectors and by abstracting a set of related components and connectors into a higher level component. Thus the top-down approach can be viewed as the inverse process of the bottom-up approach. Often both the top-down approach and the bottom-up approach have to be used together to develop a software architecture specification.

Heuristic 3 - How to Refine an Element Specification <*S, B*>

Step 1 – Refining B:

A behavior model B may be refined in several ways, for example, structure driven refinement, in which several sub-components and their connectors are identified, or functionality driven refinement, in which several functional units can be identified. Although, we do not exactly know what refinement approaches are effective in general. One thing is for sure, i.e. the input and output ports of the element must be maintained at a lower level. Petri net specific heuristics [HY92, HL91] may be used to maintain the validity of resulting lower level B'. If only behavior-preserving transformations are used to obtain B' from B, we can assure the correctness of <S, B'> based on the correctness of <S, B>; otherwise new analysis is needed to ensure the satisfiability of S [He98].

Step 2 – Refining S:

Refining S into S' in general indicates the change of requirements (a special case is when S is logically equivalent to S'), and thus results in the change of B. Once S' is known, the new B' can be developed using the approach for developing element level specification. Not any S' can be taken as a refinement of S. We require that S' maintain S, which can be elegantly expressed as S' $\Rightarrow$ S [3]. Simple heuristics such as strengthening S always result in a valid refinement S'.

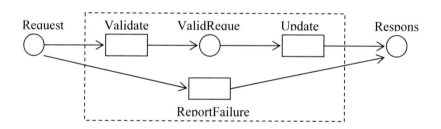

Figure 6-4. A Refined PrT Model of Transactions

As an example, Figure 6-4 shows a possible refinement of transaction into two possible scenarios in the dotted box, one is for valid request and the other for invalid request. A corresponding refinement of the property specification is

$\forall(req).(\square(Request(req) \wedge req \in Valid \Rightarrow \Diamond Response(\mathbf{S}))) \wedge$

$\forall(req).(\square(Request(req) \wedge req \notin Valid \Rightarrow \Diamond Response(\mathbf{F})))$

Where **S** and **F** stand for success and failure respectively. This refinement implies the original property specification and is thus a correct refinement according to Heuristic 3.

Heuristic 4 - How to Compose Two Element Specifications $<S_1, B_1>$ and $<S_2, B_2>$

In SAM, only a pair of related component and connector can be composed meaningfully.
Step 1 - compose B_1 and B_2 by merging identical ports;
Step 2 - compose S_1 and S_2 by conjoining $S_1 \wedge S_2$.
The soundness of viewing specification composition as logical conjunction has been shown by several researchers [AL93, ZJ93].

If we view the two transaction types, Checkout and Return, in the preceding library example as two separate components, then Figure 3 (c) illustrates the application of Heuristic 4.

6.3.3 Specify Element Instances

An element specification $<S, B>$ obtained above is generic when the initial marking in B is ignored. In PrT net, instances sharing the same net structure are distinguished through token identifications. Thus to obtain concrete elements, we only need to provide specific initial marking and generalize transition constraints to differentiate tokens with unique identifications. In general, there is no need to change the property specification S. For example, let B_1, B_2, and B_3 be three PrT nets with the same net structure and net inscription except the initial markings; then $<S, B_1>$, $<S, B_2>$, and $<S, B_3>$ are three element specifications. The above view shows the expressive power of PrT nets and first order temporal logic over that of low-level Petri nets and propositional temporal logic.

6.4 Formal Software Architecture Analysis

6.4.1 Formal Analysis Techniques

A SAM architecture description is well-defined if the ports of a component are preserved (contained) in the set of exterior ports of its refinement and the proposition symbols used in a property specification are ports of the relevant behavior model(s). The correctness of a SAM architecture description is defined by the following criteria:

(1) Element (Component / Connector) Correctness – the property specification S_{ij} holds in the corresponding behavior model B_{ij}, i.e. $B_{ij} \models S_{ij}$. Note we use B_{ij} here to denote the set of behaviors or execution sequences defined by B_{ij};

(2) Composition Correctness – the conjunction of all constraints in Cs_i of C_i is implied by the conjunction of all the property specifications S_{ij} of C_{ij}, i.e. $\wedge S_{ij} \vdash \wedge Cs_i$. An alternative weaker but acceptable criterion is that the conjunction of all constraints in Cs_i holds in the integrated behavior model B_i of composition C_i; i.e. $B_i \models \wedge Cs_i$;

(3) Refinement Correctness – the property specification S_{ij} of a component C_{ij} must be implied by the composition constraints Cs_l of its refinement C_l with $C_l = h(C_{ij})$, i.e. $\wedge Cs_l \vdash S_{ij}$. An alternative weaker but acceptable criterion is that S_{ij} holds in the integrated lower level behavior model B_l of C_l, i.e. $B_l \models S_{ij}$.

The refinement correctness is equivalent to the composition correctness when the property specification S_{ij} is inherited without change as the composition constraint Cs_l of its refinement $C_l = h(C_{ij})$. The above correctness criteria are the verification requirements of a SAM architecture description.

To ensure the correctness of a software architecture specification in SAM, we have to show that all the constraints are satisfied by the corresponding behavior models. The verification of all three correctness criteria given can be done by demonstrating that a property specification S holds in a behavior model B and, i.e. $B \models S$. The structure of SAM architecture specifications and the underlying formal methods of SAM nicely support an incremental formal analysis methodology such that the verification of above correctness criteria can be done hierarchically (vertically) and compositionally (horizontally).

Two well-established approaches to verification are model checking and theorem proving.

- *Model checking* is a technique that relies on building a finite model of a system and checking that a desired property holds in that model. Roughly speaking, the check is performed as an exhaustive state space search that is guaranteed to terminate since the model is finite. The technical challenge in model checking is in devising algorithms and data structures that allow us to handle large search spaces. Model checking has been used primarily in hardware and protocol verification [CK96]; the current trend is to apply this technique to analyzing specifications of software systems.

- *Theorem proving* is a technique by which both the system and its desired properties are expressed as formulas in some mathematical logic. This logic is given by a *formal system*, which defines a set of axioms and a set of inference rules. Theorem proving is the process of finding a proof of a property from the axioms of the system. Steps in the proof appeal to the axioms and rules, and possibly derived definitions and intermediate lemmas. Although proofs can be constructed by hand, here we focus only on machine-assisted theorem proving. Theorem provers are increasingly being used today in the mechanical verification of safety-critical properties of hardware and software designs.

6.4.2   Element Level Analysis

For each $<S_{ij}, B_{ij}>$ in composition C_i, we need to show that B_{ij} satisfies S_{ij}, i.e. $B_{ij} \models S_{ij}$. Both model checking and theorem proving techniques are applicable to element level analysis. In the following, we briefly introduce model checking technique by reachability tree [Mur89], and theorem proving technique by temporal logic [He95, He01].

(1)   Model Checking

A reachability tree is an unfolding of a PrT net, which explicitly enumerates all possible markings or states that the behavior model B_{ij} generates. The nodes of a reachability tree are reachable markings and directed edges represent feasible transitions [Mur89]. The main advantage of reachability tree technique is that the tree can be automatically generated. Once the tree is generated, different system properties can be analyzed. The main problem is space explosion when a PrT net has too many reachable states or even infinite reachable states. One possible way to deal with the above problem is to truncate the tree whenever a marking is covered by a new marking and this results in a variant of reachability trees called coverability trees. In this case, information loss is unavoidable. Thus this technique may not work in some cases. The following heuristic provides some guidelines to use the reachability tree analysis technique.

The basic idea of model checking technique for element level analysis is: (1) generating a reachability tree from B_{ij}, (2) evaluating S_{ij} using the generated reachability or coverability tree. It should be noted that when a formula contains an always operator $\square$, the formula needs to be evaluated in all nodes of the tree before a conclusion can be made.

As an example, we use the simple library system given in Figure 4 with the assumption of one valid token *req1* and one invalid token *req2* in place Request. When transition Update receives a valid request, it updates the user and book records, and generates a response *S* denoting success. When transition ReportFailure receives an invalid request, it produces a failure message *F*. The resulting reachability tree of Step (1) is shown in Figure 6-5.

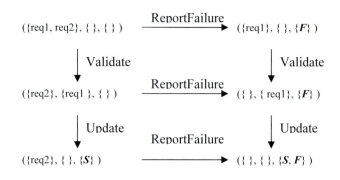

Figure 6-5. – The Reachability Tree of Figure 4

Based on Step (2), it is easy to see that the following property specification

$$\forall(req).(\Box(\text{Request}(req) \land req \in \text{Valid} \Rightarrow \Diamond\text{Response}(S)))$$

is satisfied in the reachability tree by all three possible paths: Validate – Update, ReportFailure – Validate – Update, and Validate – ReportFailure – Update. Similarly, we can evaluate the following property specification:

$$\forall(req).(\Box(\text{Request}(req) \land req \notin \text{Valid} \Rightarrow \Diamond\text{Response}(F))).$$

(2)Theorem Proving

The basic idea is to axiomatize B_{ij} [HL90, HD92] and then use the obtained axiom system to prove S_{ij}, i.e. Axiom(B_{ij}) $\vdash$ S_{ij}. The axiom system consists of general system independent axioms and inference rules and system dependent axioms and inference rules [MP83]. Each transition in B_{ij} generates a system dependent temporal logic rule that captures the causal relationships between the input places and output places of the transition. The canonical form of system dependent inference rules has the form: *fired*(t/M) $\Rightarrow$ *enabled*(t/M), where t is a transition, M is a given marking. *Fired* and *Enabled* are two predicates representing the post-condition and precondition of t under M respectively. The advantage of this technique is that a syntactic approach rather than a semantic approach is used in verification. Since no explicit representation of states is needed, there is no space explosion problem as in the reachability tree technique. The main problems are that the technique is often difficult to automate and its application requires substantial knowledge of first order temporal logic and general knowledge of theorem proof.

To demonstrate the application of this Heuristic, we axiomatize the net structure in Figure 4, and the resulting system dependent inference rules after Step1 are:

(1) $\neg M[|\text{ValidRequest}(x)|] \wedge M'[|\text{ValidRequest}(x)|] \Rightarrow M[|\text{Request}(x)|] \wedge M[|R(\text{Validate})|]$

(2) $\neg M[|\text{Response}(S)|] \wedge M'[|\text{Response}(S)|] \Rightarrow M[|\text{ValidRequest}(x)|] \wedge M[|R(\text{Update})|]$

(3) $\neg M[|\text{Response}(F)|] \wedge M'[|\text{Response}(F)|] \Rightarrow M[|\text{Request}(x)|] \wedge M[|R(\text{ReportFailure})|]$

In the above inference rules, M and M' stand for a given marking and its successor marking, respectively. $R(t)$ is the constraint associated with transition t. To prove property specification

$$\forall(req).(\square(\text{Request}(req) \wedge req \in \text{Valid} \Rightarrow \Diamond\text{Response}(S)))$$

We instantiate $\Diamond$ to a marking M' and apply rule (2) to obtain $M[|\text{ValidRequest}(x)|]$, and we apply rule (1) to obtain $M[|\text{Request}(x)|]$. With some simple logical manipulations, we can easily deduce the required property.

6.4.3 Composition Analysis

We need to show that the connected behavior model B_i (again a PrT net) of composition C_i obtained from all the individual behavior models B_{ij} ($j = 1,...,k$) of components and connectors satisfies all the constraints $\bigwedge_{c \in Cs_i} c$ in Cs_i, i.e. $B_i \models \bigwedge_{c \in Cs_i} c$. Due to the SAM framework, the analysis techniques at element level can be directly applied here. This global approach works in general, but may not be efficient.

An ideal approach is to carry out the composition level analysis compositionally. In this approach, we first analyze components and connectors individually, i.e. $B_{ij} \models S_{ij}$ for all components and connectors in a composition C_i, and then synthesize the properties, i.e. $\bigwedge S_{ij} \vdash \bigwedge_{c \in Cs_i} c$. Despite some existing results on compositional verification techniques in temporal logic [AL93] and Petri nets [HL91], their general use and application to SAM are not ready yet.

Following is a modest yet effective incremental analysis approach.

Step 1: Identify partial order relationships among the components and connectors based on their causal relationships;

Step 2: Compose and analyze the components and connectors in a partial order incrementally, starting from the least element (most independent);

Step 3: Compose and analyze mutually dependent components and connectors together;

Step 4: Once we have shown that the initial condition or marking used to prove every individual element can be ensured by the composed behavior model, then we can conclude that all the property specifications hold simultaneously.

To illustrate the ideas of this approach, let us view the refined PrT model of transactions in Figure 6-4 as a composition, which consists of three trivial components Request, ValidRequest, and Response, and three trivial connectors Validate, Update, and ReportFailure. Based on the PrT net structure, we can identify the following incremental analysis order:

(1) (Request, Validate, ValidRequest),
(2) (ValidRequest, Update, Response),
(3) (Request, Validate, ValidRequest, Update, Response),
(4) (Request, ReportFailure, Response).

Where (4) is independent of the first three analyses.

To further improve the effectiveness of this approach, we are working on some Petri net reduction techniques such that the behavior models used in incremental analysis are simplified versions of the original behavior models.

6.4.4   Refinement Analysis

For each component $C_{ij} = <S_{ij}, B_{ij}>$ with $h(C_{ij}) = C_l$, we need to show that either the connected behavior model B_l of composition C_l satisfies S_{ij}, i.e. $B_l \models S_{ij}$ or alternatively $\wedge Cs_l \models S_{ij}$. Three techniques discussed in element analysis can be used to show $B_l \models S_{ij}$. Formal temporal deduction technique [HD92, He95] can be used to prove $\wedge Cs_l \models S_{ij}$.

As an example, if we view Figure 4 as a refinement of Figure 3(d). We can easily prove the following to assure the correctness of the refinement:

$$\forall(req).(\Box(\text{Request}(req) \ \wedge \ req \in \text{Valid} \Rightarrow \Diamond\text{Response}(S))) \ \wedge$$
$$\forall(req).(\Box(\text{Request}(req) \ \wedge \ req \notin \text{Valid} \Rightarrow \Diamond\text{Response}(F))) \models$$
$$\forall(req).(\Box(\text{Request}(req) \Rightarrow \Diamond\text{Response}(msg))).$$

6.4.5   Studying Dependability Attributes Using SAM

We have studied a variety of functional properties and several non-functional dependability attributes at software architecture level using a software architecture model called SAM ([WHD99], [HD02]). We have applied SAM to specify and analyze schedulability [XHD02], performance including end-to-end latency ([WHD99], [YHG02], [SH03a]), security [HD02], fault-tolerance [SH02], reliability ([SH03a], [SH03b]) and many other functional behavior properties such as deadlock and response ([HDD02], [SH02], [HYS03]).

We have used SAM in specifying and verifying several non-functional dependability attributes including performance (end-to-end latency) [WDH99], schedulability [YHD02], security [HD02], fault tolerance [SH02], and reliability ([SH03a], [SH03b]). Since several Petri net models and temporal logics as well as a variety of formal analysis techniques were used to specify and verify the above system architectures and dependability attributes. Here we just briefly mentioned our approach without providing technical details.

- End-to-End Latency
  In [WDH99], time Petri nets [BD91] and real-time computational tree logic (CTL) [EMS92] were used to specify the software architecture of a control and command system. End-to-end latency was then verified by generating a reachability tree from the time Petri net model and evaluating timing properties specified in real-time CTL formulas. We also used stochastic Petri nets to study latency [SH03a].
- Schedulability
  In [YHG02], predicate transition nets (PrT nets) [Mur89] and first-order linear-time temporal logic (FOLTTL) ([MP92], [MP95]) were used to specify the software architecture of a simplified multi-media system. Timing requirements were dealt with by adding a time stamp attribute in tokens and by adding lower and upper bounds in transition constraints in predicate transition nets. Timing properties were specified in first-order temporal logic formulas by an additional clock variable. Verification of schedulability was again done using the theorem prover STeP.
- Security
  In [HD02], PrT nets and FOLTTL were used to specify the software architecture of an access authorization subsystem. Several system components were explicitly modeled to handle security check process. Security policies were defined as part of transition constraints within these security-checking components. Security related properties were specified using FOLTTL. Verification of security properties was done using reachability tree technique at the component level and using theorem proving at the composition level.
- Fault-Tolerance
  In [SH02], PrT nets and FOLTTL were used to specify the software architecture a communication protocol. To handle possible communication faults such as loss of information, additional system timer components were introduced to detect such losses. Fault-related properties were specified using FOLTTL and were verified using the symbolic model checker SMV [McM93].

- Reliability

In ([SH03a], [SH03b]), PrT nets were used to model a software architecture. PrT nets were then unfolded into stochastic reward nets (SRNs). Probabilistic real-time Computation Tree Logic (PCTL) [HJ94] was used to specify system reliability. The probability of system failure was then calculated using tool SPNP [Tri99] in [SH03a] and tool SMART [CJM02] in [SH03b].

6.5 Related Work

Many formal methods have been developed and applied to specifying and verifying complex software systems. For example, Z [Spi92] was used to specify software architecture [AAG95], CSP [Hoa85] was used as the foundation of Wright [AG97], and CHAM [IW95] (an operational formalism) was proposed to specify software architectures. Rapide [LKA95] used a multiple language approach in specifying software architectures, while some language has a well-defined formal foundation (for example the specification language uses a combination of algebraic and pattern constraints), others offer constructs similar to those in a typical high-level programming language.

Two complementary formal methods, Petri nets and temporal logic, are used in SAM to define behavior models and property specifications respectively. The selection of the above formal methods is based on the following reasons. Well-known model-oriented formal methods include Petri nets and finite state machines. Finite state machines are simple, but have difficulty to deal with concurrent systems especially distributed systems. Petri nets are well suited for modeling concurrent and distributed systems, which characterize the majority of embedded systems being used by NASA and other government agencies. However Petri nets are often misunderstood and even prejudiced in the U.S. Many researchers' knowledge of Petri nets is limited to the 1st generation low-level Petri nets used primarily for modeling control flows. Petri nets have evolved tremendously in the past 20 years, from the 2nd generation high-level Petri nets in 1980s ([JR91]) and the 3rd generation hierarchical and modular Petri nets in early 1990s ([HL91], [He96], [Jen92]) to the 4th generation object-oriented Petri nets in late 1990s [ADR01]. More importantly, Petri nets have been extended in many different ways to study system performance, reliability, and schedulability ([MBC94], [Wan98], [Haa02]); which are the central attributes of complex dependable systems. There are vast existing research results on Petri nets (over 8000 publications). Despite many different types of temporal logic, for example, propositional vs. first-order, linear time vs. branch time, timed vs. un-timed, probabilistic vs. non-probabilistic, it is widely accepted that temporal logic in general is an excellent property-oriented formal method for specifying behavioral properties of concurrent systems. We are familiar with and have extensive experience in using Manna & Pnueli's linear-time first order temporal logic ([MP92], [MP95]), Lamport's linear-time first order temporal logic (Temporal Logic of Actions) [Lam94], and Clarke and Emerson's branch time propositional logic CTL [CE81] and its extension CTL* [CES86]; and various timed versions of the above temporal logics ([AH92], [AM94], [EMS92]). One major problem of using a dual-formalism is how to integrate two formal methods in a consistent and meaningful way, our own research results ([HL90], [HD92], [He92]) and others work [MMP96] have provided a satisfactory solution to integrate Petri nets and temporal logic in SAM.

Almost all ADLs support the specification and analysis of major system functional properties such as safety and liveness properties [MT00]. Several ADLs also provide capabilities to represent some dependability attributes. MetaH [BEJ96] supported the description of non-functional properties such as real-time schedulability, reliability and security in components but not in connectors. Unicon [SDK95] supported the definition of real-time schedulability in both components and connectors. Rapide [LKA95] supported the modeling of time constraints in architectural configurations. The analysis of non-functional properties in the above ADLs was not performed at the architecture specification level instead of during the simulation and implementation. As pointed out in [SR98] "ADLs need to be extended with appropriate linguistic support for expressing dependability constraints. They also need to be furnished with an appropriate semantics, to enable formal verification of architectural properties."

6.6 Concluding Remarks

Commercial pressure to produce higher quality software is always increasing. Formal methods have already demonstrated success in specifying commercial and safety-critical software, and in verifying protocol standards and hardware designs. In this chapter, we have provided a well-defined integration of two well-known formal methods predicate transition nets and first order linear-time temporal logic as the foundation for writing software architecture specifications in SAM. This dual formal methods approach supports both behavioral modeling and property analysis of software architectures. Unlike many other architecture description language research efforts that primarily focus on the representation issues of software architectures, we have further presented a unified framework with a set of heuristics to develop and analyze software architecture specifications in SAM. The heuristics are supported by well-developed existing techniques and methods, with potential software tool assistance. We have demonstrated the applications of several of the heuristics with regard to the development and analysis in a non-trivial example. Our contributions are not limited to software architecture research, but also shed some light on how mature formal methods can be effectively used in real-world software development. While it is true that every formal method has its limits and weaknesses, it is important to rely on its strengths while avoiding and minimizing its weaknesses in practical applications. The above philosophy has been used both in designing our dual formal methods foundation of SAM as well as our framework consisting of a variety of development and analysis techniques.

SAM has been applied to model and analyze the software architectures of several systems, including a control and command system [WHD99], a flexible manufacturing system [WD99], popular architectural connectors [HD00], the alternating bit communication protocol, and a resource access decision system here. We are carrying out more case studies to explore the effectiveness of combining different development and analysis techniques and to determine the practical limitations of each individual technique. To support this whole SAM framework, we are adding software components to our existing SAM environment, which consists of a graphical editor for building behavioral models, a textual editor for defining property specifications, a simulator to execute behavioral models, and an analyzer to model check the property specifications in the behavioral models.

6.7 Acknowledgements

This research was supported in part by the National Science Foundation of the USA under grant HRD-0317692, and by the National Aeronautics and Space Administration of the USA under grant NAG2-1440.

6.8 References

[AAG95] G. Abowd, R. Allen, and D. Garlan: "Formalizing Style to Understand Descriptions of Software Architecture", *ACM Transaction on Software Engineering and Methodology*, vol.4, no.4, 1995.

[AG97] R. Allen, and D. Garlan: "A formal Basis for Architectural Connection", *ACM Transaction on Software Engineering and Methodology*, vol.6, no.3, 1997, 213-249.

[AL91] M. Abadi and L. Lamport: "The existence of refinement mappings", *Theoretical Computer Science*, 82, 1991, 253-284.

[AL93] M. Abadi and L. Lamport: "Composing specification", *ACM Trans. on Programming Languages and Systems*, Vol. 15, 1993, 73-130.

[BD91] B. Berthomieu and M. Diaz, Modeling and Verification of Time Dependent Systems Using Time Petri Nets, *IEEE Trans. Software Engineering*, Vol. 17, No. 3, 1991, 259-273.

[BRJ99] G. Booch, J. Rumbaugh, and I. Jacobson: *The Unified Modeling Language – User Guide*, Addison Wesley, 1999.

[CE81] E.M. Clarke and E.A. Emerson, "Characterizing Properties of Parallel Programs as fixpoints", *Proc. of the 7th International Colloquim on Automata, Languages, and Programming, Lecture Notes in Computer Science*, vol.85, 1981.

[CES86] E.M. Clarke, E.A. Emerson, and A.P. Sistla, "Automatic Verification of Finite-State Concurrent Systems using Temporal Logic Specifications", *ACM Trans. on Programming Languages and Systems*, vol.8, no.2, 1986, 244-263.

[CK96] E. M. Clarke, O. Grumberg and D. A, Peled, *Model Checking*, The MIT Press, 1999.

[CGR95] D. Craigen, S. Gerhart, and T. Ralston: "Formal Methods Reality Check: Industrial Usage", *IEEE Trans. On Software Engineering*, vol.21, no.2, 1995.

[CJM02] G. Ciardo, R. Jones, R. Marmorstein, A. Miner, and R. Siminiceanu, "SMART: Stochastics model-checking analyzer for reliability and timing", *Proc. of Int'l Conf. on Dependable Systems and Networks*, Washington, June 2002.

[CK96] E. M. Clarke and R. Kurshan, "Computer-aided verification", *IEEE Spectrum*, Vol.33, No.6 , 1996, 61-67.

[CW96] E. Clarke and J. Wing: "Formal Methods: State of the Art and Future", *ACM Computing Surveys*, vol.28, no.4, 1996, 626-643.

[DWB03] Y. Deng, J. Wang, K. Beznosov and J. J.P. Tsai, "An approach for modeling and analysis of security system architectures", *IEEE Transactions on Knowledge and Data Engineering*, Vol. 15, No. 2, March/April 2003.

[EMS92] E. Emerson, A. Mok, A. Sistla, and J. Srinivasian: "Quantitative Temporal Reasoning", *Real-Time Systems*, vol.4, 1992, 331-352.

[Haa02] P. Haas: *Stochastic Petri Nets: Modeling, Stability, Simulation,* Springer-Verlag, 2002.

[Har87] D. Harel: Statecharts: a visual formalism for complex systems. *Science of Computer Programming,* vol. 8, 231-274.

[HD92] X. He and Y. Ding: "A Temporal Logic Approach for Analyzing Safety Properties of Predicate Transition Nets", *Proc. of the 12th IFIP World Computer Congress (Information Processing'92),* Madrid, Spain, 1992, 127-133.

[HD00] X. He and Y. Deng: " Specifying software architectural connectors in SAM", *International Journal of Software Engineering and Knowledge Engineering,* 10, 2000, 411-432.

[HD02] X. He and Y. Deng: "A Framework for Developing and Analyzing Software Architecture Specifications in SAM", *The Computer Journal*, vol.45, no.1, 2002, 111-128.

[HDD02] X. He, J. Ding, and Y. Deng: "Analyzing SAM Architectural Specifications Using Model Checking", *Proc. of SEKE2002*, Italy, 2002.

[He92] X. He: "Temporal Predicate Transition Nets - A New Formalism for Specifying and Verifying Concurrent Systems", *International Journal of Computer Mathematics*, vol.45, no.1/2, 1992, 171-184.

[He95] X. He: "A method for analyzing properties of hierarchical predicate transition nets", *Proc. of the 19th Annual International Computer Software and Applications Conference (COMPSAC'95)*, Dallas, Texas, August, IEEE Computer Society Press, U.S.A., 1995, 50-55

[He96] X. He: "A Formal Definition of Hierarchical Predicate Transition Nets", *Proc. of the 17th International Conference on Application and Theory of Petri Nets (ICATPN'96)*, Lecture Notes in Computer Science, vol. 1091, Osaka, Japan, 1996, 212-229.

[He98] X. He: "Transformations on hierarchical predicate transition nets: abstractions and refinements", *Proc. of the 22nd International Computer Software and Application Conference (COMPSAC'98)*, Vienna, Austria, August , IEEE Computer Society Press, U.S.A., 1998, 164-169

[He01] X. He: "PZ nets - a formal method integrating Petri nets with Z", *Information and Software Technology,* 43, 2001, 1-18.

[HJ94] H. Hansson and B. Johnson, "A Logic for Reasoning about Time and reliability", *Formal Aspects of Computing*, vol.6, no.4, 1994, 512-535.

[HL90] X. He and J.A.N. Lee, "Integrating predicate transition nets and first order temporal logic in the specification of concurrent systems", *Formal Aspects of Computing*, vol.2, no.3, 1990, 226-246.

[HL91] X. He and J.A.N. Lee, "A methodology for constructing predicate transition net specifications", *Software - Practice & Experience*, 21, 1991, 845-875.

[Hoa85] C.A.R. Hoare, *Communicating Sequential Processes*, Prentice-Hall, 1985.

[HY92] X. He and C.H. Yang: "Structured analysis using hierarchical predicate transition nets" , *Proc. of the 16th Int'l Computer Software and Applications Conf. (COMPSAC'92)*, Chicago, September, IEEE Computer Society Press, U.S.A., 1992, 212-217.

[KKC00] R. Kazman, M. Klein, P. Clements: "ATAM: A Method for Architectural Evaluation" *Software Engineering Institute Technical Report* CMU/SEI-2000-TR-004.

[Kni02] J. Knight: "Dependability of Embedded Systems", *Proc. of ICSE'02*, Orlando, 2002, 685-686.

[Jen92] K. Jensen, K. *Coloured Petri Nets*, Springer-Verlag, Berlin, 1992.

[Lam94] L. Lamport: "The Temporal Logic of Actions", *ACM Transactions on Programming Languages and Systems*, vol.16, no.3, 1994, 872-923.

[LKA95] D.C. Luckham, J. Kenney, L. Augustin et al: Specification and Analysis of System Architecture Using Rapide, *IEEE Transaction on Software Engineering*, vol.21, no.4, 1995, 336-355.

[MBC94] M. Marsan, G. Balbo, G. Conte, S. Donatelli, G. Franceschinis: *Modeling with Generalized Stochastic Petri Nets*, John Wiley and Sons, 1994.

[McM93] K. L. McMillan, *Symbolic Model Checking*, Kluwer Academic Publishers, Boston, 1993.

[Mil89] R. Milner, *Communication and Concurrency*, Prentice-Hall, 1989.

[MMP96] D. Mandrioli, A. Morzenti, M. Pezze, P. Pietro S. and S. Silva. A Petri net and logic approach to the specification and verification of real time systems. *Formal Methods for Real time Computing*, 1996.

[MP83] Z. Manna and A. Pnueli: "How to cook a temporal proof system for your pet language", *Proc. Of the 10th ACM Symp. On Principle of Programming Languages*, Austin, Texas, ACM Press, 1983, 141-154

[MP92] Z. Manna and A. Pnueli: *The Temporal Logic of Reactive and Concurrent Systems - Specification*, Springer-Verlag, 1992.

[MP95] Z. Manna and A. Pnueli: *The Temporal Verification of Reactive Systems - Safety*, Springer-Verlag, 1995.

[MT00] N. Medvidovic and R. Taylor: "A Classification and Comparison Framework for Software Architecture Description Languages", *IEEE Transaction on Software Engineering*, vol.26, no.1, 2000, 70-93.

[Mur89] T. Murata: " Petri nets, Properties, analysis and applications", *Proc. of IEEE*, vol.77, no.4, 1989, 541-580.

[Rei92] W. Reisig. *A Primer in Petri Net Design*, Springer-Verlag, Berlin, 1992.

[SDK95] M. Shaw, R. Deline, D. Klein et al: Abstractions for Software Architecture and Tools to Support Them, *IEEE Trans. on Software Eng.*, vol. 21, no.4, 1995, 314-335.

[SG96] M. Shaw and D. Garlan, Software Architecture, Prentice-Hall, 1996.

[SH02] T. Shi and X. He: "Modeling and Analyzing the Software Architecture of A Communication Protocol Using SAM", *Proc. of the 3rd Working IEEE/IFIP Conference on Software Architecture*, Montreal, August, 2002.

[SH03a] T. Shi and X. He: "Dependeability Analysis using SAM", *Proc. of the ICSE Workshop on Software Architectures for Dependable Systems*, May 3, Portland, Oregon, USA, 2003.1

[SH03b] T. Shi and X. He: "A Methodology for Dependability and Performability Analysis in SAM", *Proc. of The International Conference on Dependable Systems and Networks*, San Francisco, CA, June 22nd - 25th, 2003.

[Spi92] Spivey, Z Reference Manual, Prentice-Hall, 1992.

[SR98] V. Stavridou and R. Riemenschneider: "Provably Dependeable Software Architectures", *Proc. of 3rd International Software Architecture Workshop*, Florida, 1998, 133-136.

[Tri99] K. Trivedi, *SPNP User's Manual*, version 6.0, Department of ECE, Duke University, 1999.

[Wan98] J. Wang: *Timed Petri Nets, Theory and Application*, Kluwer Academic Publisher, 1998.

[WHD99] J. Wang, X. He, Y. Deng: "Introducing Software Architecture Specification and Analysis in SAM through An Example", *Information and Software Technology*, vol.41, 1999, 451-467.

[Win90] J. Wing: "A Specifier's Introduction to Formal Methods", *IEEE Computer*, 23(9): 8-24, 1990.

[XHD02] D. Xu, X. He, and Y. Deng: "Compositional Schedulability Analysis of Real-Time Systems Using Time Petri Nets", *IEEE Trans. On Software Engineering*, vol.28, no.10, 2002.

[YHD02] H. Yu, X. He, S. Gao and Y. Deng: " Modeling and Analyzing SMIL Documents in SAM", *Proc. of Fourth IEEE International Symposium on Multimedia Software Engineering,* Newport Beach, California, USA, pp.132-139, 2002

[ZJ93] P, Zave and M. Jackson: "Conjunction as composition", *ACM Transaction on Software Engineering and Methodology*, 2, 1993, 379-411.

7 An Algebraic Approach to Hardware Compilation

Authors

Jonathan P. Bowen[1]   and   He Jifeng[2]

[1] London South Bank University
Centre for Applied Formal Methods, Institute for Computing Research
Faculty of Business, Computing and Information Management, UK

[2] The United Nations University
International Institute for Software Technology, China

Summary

This chapter presents a provably correct compilation scheme that converts a program into a network of abstract components that interact with each other by exchanging request and acknowledgement signals. We provide a systematic and modular technique for correctly realizing the abstract components in hardware device, and use a standard programming language to describe both algorithms and circuits. The resulting circuitry, which behaves according to the program, has the same structure as the program. The circuit logic is asynchronous, with no global clock.

Keywords: compilation; formal methods; hardware design

7.1 Introduction

With chip sizes consisting of millions of transistors, the complexity of VLSI algorithms – i.e., algorithms implemented as a digital VLSI circuits – is approaching that of software algorithms – i.e., algorithms implemented as code for a stored-program computer. For many applications, particularly where speed of execution or security is important, a customer-built circuit is better than the traditional processor-and-software combination. The speed is improved by the absence of the machine language layer and also by introducing parallelism, whereas security is improved by the impossibility of reprogramming. Moreover, there are spacing saving compared to a combination of software and processor.

Hossam A. Gabbar (ed.), Modern Formal Methods and Applications, 151–176.
© 2006 *Springer. Printed in the Netherlands.*

Yet design methods for VLSI algorithms lag far behind the potential of the technology. The design methods for digital circuits that are commonly found in textbooks resemble the low-level machine-language programming methods. Selecting individual logic gates in a circuit is something like selecting individual machine instruction in a program. These methods may have been adequate for small circuit design when they were introduced, and they may still be adequate for large circuits that are simply repetitions of a small circuit (such as a memory), but they are not adequate for circuits that perform complicated customer algorithm.

Since a VLSI system is a highly concurrent computation, we propose an approach to VLSI design based on concurrent computing. The circuit to be designed is first implemented as a concurrent program that fulfils the logical specification of the circuit. The program is then compiled into a circuit by applying semantics-preserving transformations. Hence, the circuit obtained is correct by construction.

Communication in VLSI is becoming increasingly expensive, compared to switching, as the size of the wire determines both the switching costs and the area of a chip. In order to reflect those cost ratios, a model in which communication is explicit is more appropriate to control the cost of communication. Hence, we opted for a notation based on the notion of concurrent processes communicating by explicit message-passing and assignments to variables. We adopt a high level programming language like Occam [7] as a behavior specification language for hardware device. Occam is a language for designing and describing concurrent systems, and hardware designers exploit concurrency in their pursuit of increased performance. For example, today's fastest microprocessors typically contain a number of cooperating agents: a bus interface that directs traffic among the main memory and caches; an instruction fetch unit that reads and decodes instructions from the cache; and several execution units that carry out the decoded instructions. The components of the microprocessor synchronize when they need to exchange information, but otherwise proceed at their own pace. Such a system is naturally described as a set of communicating processes.

This chapter presents a *provably correct* compilation scheme that converts a program into a network of abstract components that interact with each other by exchanging *request* and *acknowledgement* signals. We provide a systematic and modular technique for correctly realizing the abstract components in hardware device, and use a standard programming language to describe both algorithms and circuits. The resulting circuits, which behave according to the program, have the same *structure* as the program. The circuit logic is asynchronous, with no global clock.

Why is it significant that our compilation scheme is verified? Highly concurrent systems are notoriously difficult implement correctly; there is little chance of getting them right unless a disciplined approach is taken early in their specification and design. Occam's concise notation makes it easy to see whether a given description captures the designer's intent. Furthermore, Occam has a well-understood mathematical model [4] and a complete set of algebraic laws [13], allowing potential system misbehavior to be detected by analysis rather than simulation. However, rigorous reasoning at the source level is for naught if the compilation scheme itself introduces misbehavior. The importance of detecting flaws before a product goes to market was understood by the Pentium disaster of 1994: Intel was forced to write off $475 million due to an obscure bug in the Pentium's floating-point division unit [8].

The VHDL [10] and Verilog [15] languages are presently being used by industry. They provide a way to express formally and symbolically the constituent components of a hardware circuit and their interconnections, and allow circuits designers to describe circuits more conveniently, but they are not translated automatically to circuits. There are interactive synthesis tools to aid in the construction of synchronous circuits from a subset of these languages. The circuits are then *verified* by simulation.

There are other high-level circuit design methods that have been developed and reported in the literature. Martin at CalTech developed a method of compiling a communicating process [6] into a set of transistors via an intermediate mapping to production rules. In [1, 2], a similar approach (and a similar circuit design language) was taken, except that specifications are mapped into connections of small components for which standard transistor implementations exist. In [16], circuits are modeled as networks of finite state machines, using their formalism to assist in proving the correctness of their compiled circuits. Page at Oxford developed a prototype compiler in the functional language SML, which converted an Occam-like language to a *netlist* [1]. After further processing by vendor software the netlist can be loaded into Xilinx FPGA chips [17]. This work is most similar to ours, but their designs have a global clock; ours do not. Moreover, the algebraic approach in this chapter offers the significant advantages of providing a provably compiling method, and it is also expected to support a wide range of design optimization strategies.

The rest of this chapter is organized as follows. Section 7.2 presents a simple language of communicating processes as the source language. An overview of our compilation strategy is given in Section 7.3. Section 7.4 introduces the concepts of handshake protocol and context-dependent refinement. Section 7.5 is devoted to the implementation of program variables and Boolean expressions. In Section 7.6, we convert the sequential subset of the language into a network of abstract components, and validate the compilation scheme. Section 7.7 shows how to use a set of primitive circuits to implement the control processes generated by our hardware compilation scheme.

7.2 A Language of Communicating Processes

7.2.1 Syntax

Our language of communicating processes contains the features that are typical of Occam and Occam-like languages:

- Boolean state variables can be used in expressions and updated by assignment.
- Programs can be composed in sequence, in parallel, or made to execute conditionally or repeatedly.
- Concurrently executing programs can synchronize through shared objects called channels.
- Programs offer several channels on which to synchronize may choose among them.

The syntax for the language is given by the following BNF rules, where x stands for a program variable of Boolean type, ch for a channel name, b for a Boolean expression, and P for a process, and for sequence catenation.

$$P \quad ::= \quad \textbf{skip} \mid x := b \mid ch?x \mid ch!b \mid P \sqcap P \mid$$

$$P; P \mid P \| P \mid \textbf{if } BG \textbf{ fi} \mid \textbf{do } BG \textbf{ od} \mid \textbf{alt } G \textbf{ tla}$$

$$BG \quad ::= \quad <b \to P> \mid BG \frown BG$$

$$G \quad ::= \quad <ch?x \to P> \mid G \frown G$$

$$b \quad ::= \quad true \mid false \mid x \mid \neg b \mid b \wedge b \mid b \vee b$$

Informally, the process terms stand for the following processes:

1. Execution of **skip** does nothing, and leaves all variables unchanged.
2. Execution of $x := b$ assigns the value of expression b to variable x.
3. $ch?x$ is input from a channel named ch to variable x.
4. $ch!b$ is output to a channel named ch of the value of expression b.
5. The construct $P \, . \, Q$ executes either P or Q, where the choice between P and Q is made non-deterministically, without consent of its environment.

   In general, let P be a finite non-empty set of processes. We use the notation $.$P to denote the non-deterministic choice over the members of P.

   We define a relation $.$ between programs such that P holds whenever, for any purpose, the observable behavior of P is good as, or better than, that of Q:

$$P \, . \, Q =_{df} (P \, . \, Q) = Q$$

6. The sequential composition $P;Q$ executes P first, and then executes Q after P terminates; it terminates when Q terminates.

7. $P \parallel Q$ stands for the parallel composition of P and Q, wherein all communications between P and Q are concealed from the environment. Let E be the set of communication events between P and Q. The synchronization construct $P \parallel_E Q$ behaves like the parallel composition $P \parallel Q$ except that the communication events of E remain visible to the external environment

$$P \parallel Q = (P \parallel_E Q) \setminus E$$

where $\setminus$ denotes the Communicating Sequential Processes (CSP) hiding operator [6, 14].

8. The Boolean guarded process $b \to P$ cannot appear outside of a conditional or iterative statement; it is triggered only when the initial value of b is true and its execution completes when P terminates. Several guarded processes may be composed into a sequence.

9. The conditional **if** BG **fi** examines its guarded processes in order, selects the first one whose Boolean guard is true, and executes its associated process. It terminates when that process terminates. If none of its guards is true, the behavior of the conditional becomes totally unpredictable like a chaotic program.

   For notational simplicity, we will use $P . b . Q$ to describe a program which behaves like P if the initial value of b is true, or like Q if the initial value of b is false:

$$P \triangleleft b \triangleright Q =_{df} \text{ if } (b \to P, \ \neg b \to Q) \text{ fi}$$

10. The iterative construct **do** BG **od** is similar to the conditional **if** BG **fi**, except that it repeatedly evaluates its guarded processes until none of its guards is true, and then it terminates.

11. The guarded alternation **alt** G **tla** offers to its environment a choice over the input guards of its alternatives. If the environment performs a communication on c, and $c?x \to P$ is the unique element of sequence G with the input on c as the guard, then the construct **alt** G **tla** will behave like process P. If there are a number of guarded processes in G with the input on c as their guards, then the choice among them will be made non-deterministically.

    We adopt the CSP notation $P \parallel Q$ to stand for the external choice between P and Q, which offers the environment the choice of the first events of P and Q and then behaves accordingly. The alternation construct can be rewritten as external choice:

$$\textbf{alt } (c_1?x_1 \to P_1 \ \dots \ c_n?x_n \to P_n) \textbf{ tla } = (c_1?x_1 \to P_1) \| \dots \| (c_n \to P_n)$$

Let $A = (a_1,\dots,a_n\}$ be a finite set of communication events. The notation

$$x : A \to P(x)$$

abbreviates

$$(a1 \to P(a1)) \| (a2 \to P(a2)) \| \ \dots \ \| (a_n \to P(a_n))$$

12. The Boolean expressions include *true* ("high voltage" or "power") and *false* ("low voltage" or "ground"). The Boolean unary operator ¬ is negation. The Boolean binary operators include ∧ (conjunction) and ∨ (disjunction).

A local variable x can be introduced by the declaration command **var** $x \bullet P$. We will also use **chaos** to represent the chaotic program whose behavior is totally uncontrollable and unpredictable.

In the rest of this chapter, we take a more general form of recursion and use the notation $\mu X \bullet P(X)$ to stand for the *weakest fixed point* (with respect to the relation .) of the equation $X = P(X)$.

Example 2.1 (clock)
A clock can be modeled by a recursive process *CLOCK* which can repeatedly engage in the event *tick*

$$CLOCK =_{df} \mu X \bullet (tick \rightarrow X)$$

Example 2.2 (a single place buffer)
A singe place buffer, which inputs messages on channel *left* and outputs them on channel *right*, can be modeled by the recursive process *COPY*:

$$COPY =_{df} \mu X \bullet (left?x \rightarrow (right!x \rightarrow X))$$

Example 2.3 (wire)
The process *WIRE(a,c)* can emit an output event c in response to an input event a. It is also *receptive* in the sense that it does not refuse input, but may diverge if given an input that it is not prepared to handle.

$$WIRE(a, c) =_{df} \mu X \bullet (a? \rightarrow ((a? \rightarrow \mathbf{chaos}) \| (c! \rightarrow X)))$$

Example 2.4 (iteration)

The iteration $\mathbf{do}(b_1 \rightarrow P_1,...,b_n \rightarrow P_n)\mathbf{od}$ can be rewritten as the tail recursion:

$$\mu X \bullet (\mathbf{if}\ (b_1 \rightarrow P_1, \ldots, b_n \rightarrow P_n)\ \mathbf{fi};\ X) \lhd (b_1 \vee \ldots \vee b_n) \rhd \mathbf{skip}$$

7.2.2 Algebraic laws

The basic laws defining Occam-like programs are given in [13]. This section gives a number of algebraic laws that are used in the design and verification of the hardware compilation scheme presented later in this chapter.

The non-deterministic choice operator is idempotent, symmetric, associative and disjunctive. It has **chaos** as its zero.

Law 1 (non-deterministic choice)

1.1 $P . P = P$

1.2 $P . Q = Q . P$

1.3 $P . (Q . R) = (P . Q) . R$

1.4 $P . (Q . R) = (P . Q) . (P . R)$

1.5 $P .$ **chaos** $=$ **chaos**

The relation . induced by the non-deterministic choice operator is a partial order; i.e., it is reflexive, transitive and anti-symmetric. The chaotic program **chaos** is the bottom of the relation . .

Law 2 (refinement)

2.1 $P .$ **chaos**

Sequential composition is associative and disjunctive, and has unit **skip** and left zero **chaos**.

Law 3 (sequence)

3.1 $(P;Q);R = P;(Q;R)$

3.2 $(P . Q);R = (P;R) . (Q;R)$

3.3 $P; (Q . R) = (P;Q) . (P;R)$

3.4 **skip**$;Q = Q =$ $Q;$**skip**

3.5 **chaos**$;Q =$ **chaos**

Conditionals are product. It is idempotent, skew-symmetric, associative and disjunctive. Sequential composition distributes leftward over the conditional.

Law 4 (conditional)

4.1 $P .b .P = P$

4.2 $P .true .Q = P$

4.3 $P .b .Q = Q . \neg b .P$

4.4 $(P .b .Q) .c .R = P .b \wedge c . (Q .c .R)$

4.5 $P .b . (Q .c .R) = (P .b .Q) .c .(P .b .R)$

4.6 $P .b . (Q .R) = (P .b .Q) .(P .b .R)$

4.7 $(P .b . Q); R = (P;R) .b .(Q;R)$

The following law connects the "**if**" construct with conditional.

Law 5 (if)

5.1 **if** $b \rightarrow P$ **fi** $= P \cdot b \cdot$**chaos**

5.2 **if** $<b \rightarrow P>$. BG **fi** $= P \cdot b \cdot$(**if** BG **fi**)

The external choice operator $[\!]$ is idempotent, symmetric associative and disjunctive, and has **chaos** as its zero. It distributes over sequential composition when all its components are guarded.

Law 6 (external choice)

6.1 $P [\!] P = P$

6.2 $P [\!] Q = Q [\!] P$

6.3 $P [\!] (Q [\!] R) = (P [\!] Q) [\!] R$

6.4 $P [\!] (Q \cdot R) = (P [\!] Q) \cdot (P [\!] R)$

6.5 $(a \rightarrow P) [\!] (a \rightarrow Q) = a \rightarrow (P \cdot Q)$

6.6 $P [\!]$ **chaos** $=$ **chaos**

6.7 $(x{:}A \rightarrow P(x)); Q = x{:}A \rightarrow (P(x);Q)$

The parallel operators $\|$ and $\|_E$ are symmetric, associative and disjunctive, and have **chaos** as zero.

Law 7 (parallel)

7.1 $P \| Q = Q \| P$

7.2 $P \| (Q \| R) = (P \| Q) \| R$

7.3 $P \| (Q \cdot R) = (P \| Q) \cdot (P \| R)$

7.4 $P \|$ **chaos** $=$ **chaos**

When P can perform any events A, being equivalent to a process of the form $x{:}A \rightarrow P(x)$, and Q can perform the events in B, respectively $y{:}B \rightarrow Q(y)$, then $P \|_E Q$ can perform any event of

$$(A \setminus E) \cup (B \setminus E) \cup (A \cap B)$$

The first component of this union are the events P can perform on its own (because they are not in E); similarly, the second are the events Q can do by itself. The final components are the events on which they can synchronize (i.e., the ones that they both can do). The law expressing this is the following:

Law 8 (expansion law of $\|_E$)

Let $P = x{:}A \rightarrow P(x)$ and $Q = y{:}B \rightarrow Q(y)$.

8.1 $P \|_E Q = (x{:}A\backslash E \rightarrow (P(x) \|_E Q) [\!] (y{:}B\backslash E \rightarrow (P \|_E Q(y)) [\!] (z{:}A\cap B \rightarrow (P(z) \|_E Q(z))$

The expansion law of $\parallel$ is more complicated because any event in $A \cap B$ can occur silently and will be treated as an internal communication, and thus is concealed from the environment:

Law 9 (expansion law of $\parallel$)

9.1 $P \parallel Q = (x{:}A\backslash E \rightarrow (P(x) \parallel_E Q) \parallel (y{:}B\backslash E \rightarrow (P \parallel_E Q(y))$

$\parallel (._{z \in A \cap B} \rightarrow (P(z) \parallel_E Q(z))) .(._{z \in A \cap B} \rightarrow (P(z) \parallel_E Q(z)))$

Both expansions laws enable us to transfer a parallel program into a sequential one, and are widely used in the later proof.

The program $\mu X \bullet P(X)$ can be implemented as a single non-recursive call of a parameterless procedure with name X and with body $P(X)$. Occurrences of X within $P(X)$ are implemented as recursive calls on the same procedure. The following laws [15] state that $\mu X \bullet P(X)$ is indeed a fixed point of P, and that it is the weakest one.

Law 10 (weakest fixed point)

10.1 $\mu X \bullet P(X) = P(\mu X \bullet P(X))$

10.2 if $Y . P(Y)$ then $Y . \mu X \bullet P(X)$

All the recursion we have seen (such as *CLOCK*, *COPY* and *WIRE*) and all most all recursions one meets in practice) have a property that makes them easier to understand and reason about. They are *guarded*, i.e., each recursive call comes after a communication that is introduced by the recursive definition. The point about a guarded recursion is that the first-step behavior does not depend on at all on a recursive call, and when a recursive call is reached, the first step of its behavior, in turn, can be computed without any deeper call. This leads to the principle of unique fixed point for guarded recursion.

Law 11 (unique fixed point)

If $\mu X \bullet P(X)$ is a guarded recursion and Y satisfies the equation $Y = P(Y)$ then $Y = \mu X \bullet P(X)$.

The following laws express the basic properties of assignment: that variables not mentioned on the left of "$:=$" remain unchanged, that the order of the listing is immaterial, and that evaluation of an expression or a condition uses the value most recently assigned to its variables.

Law 12 (assignment)

12.1 $(x, y := e, y) = x := e$

12.2 $(x, ..., y, ... := e, ..., f, ...) = (y, ..., x, ... := f, ..., e, ...)$

12.3 $(x := e; x := f(x)) = x := f(e)$

12.4 $(x := e); (P .b(x) .Q) = (x := e; P); .b(e) .(x := e; Q)$

Let Σ be the set of events which the process P can perform. Given $L \subseteq \Sigma$ and a divergence-free process P, the process $P \backslash\backslash L$ represents an abstract view of P in the set $\Sigma \setminus L$, and is defined in [14] by:

$$Divergences(P \backslash\backslash L) =_{df} .$$

$$Failures(P \backslash\backslash L) =_{df} \{(s \downarrow(\Sigma \setminus L), X) \mid (s, X \cap (\Sigma \setminus L)) \ Failures(P)\}$$

The process projection construct $P \backslash\backslash L$ differs from the hiding construct $P \setminus L$ in the following way: in the failure-divergence model the latter will introduce a divergence whenever P can perform an infinite sequence of events of L, but the former becomes deadlock.

Law 13 (projection)

13.1 If the alphabet of P_i is disjoint from the set L_{1-i} for $i = 0, 1$, then

$$(P0 \parallel P1) \backslash\backslash (L0 \cup L1) = (P0 \backslash\backslash 0) \parallel (P1 \backslash\backslash 1)$$

13.2 If P is a process without containing the process variable X, then

$$\mu X \bullet (a \rightarrow P \parallel b \rightarrow X) \backslash\backslash \{a\} = a \rightarrow (P \backslash\backslash \{a\})$$

As an example to show how to apply algebraic laws, we are going to establish the following fact: *two wires can be connected into a single one.*

Lemma 2.5 (connecting wires)

$$WIRE(a,h) \parallel WIRE(h,c) = WIRE(a,c)$$

Proof: Let $FULL(a,c) =_{df} (a? \rightarrow \mathbf{chaos} \parallel c! \rightarrow WIRE(a,c))$. We have

$$WIRE(a,c) = a? \rightarrow FULL(a,c)$$

LHS

$=$ {**Law 9**}

$a? \to (FULL(a, h)\|WIRE(h, c))$

$=$ {**Law 9**}

$a? \to (a? \to (chaos\|WIRE(h, c)) \mathbin{\|} (WIRE(a, c)\|FULL(h, c))$

$\qquad \sqcap (WIRE(a, h)\|FULL(h, c))$

$=$ {**Laws 7.4 and 9**}

$a? \to (a? \to chaos \mathbin{\|} a? \to (FULL(a, h)\|FULL(h, c)) \mathbin{\|} c! \to LHS)$

$\qquad \sqcap (a? \to (FULL(a, h)\|FULL(h, c)) \mathbin{\|} c! \to LHS)$

$=$ {**Laws 6.5 and 1.5**}

$a? \to (a? \to chaos \mathbin{\|} c! \to LHS)$

$\qquad \sqcap (a? \to (FULL(a, h)\|FULL(h, c)) \mathbin{\|} c! \to LHS)$

$=$ {**Laws 6.4, 6.5 and 1.5**}

$a? \to (a? \to chaos\|c! \to LHS)$

which together with the unique fixed point theorem **Law 10** implies that

$$LHS = \mu X \bullet (a? \to (a? \to \mathbf{chaos} \mathbin{\|} c! \to X)) = RHS \qquad \Box$$

7.3 Compiling Strategy

The main difference between software programming and VLSI programming is that in VLSI, concurrency is free and sequencing is costly, whereas it is just the opposite in software. In hardware, concurrency is implemented by mere juxtaposition of circuits. Sequencing requires synchronization. We therefore avoid sequencing as much as possible, and implement it as a restricted form of concurrency. This is the compilation strategy adopted in this chapter.

One small technicality is that rather than implement P directly, we choose to implement a *reusable* version of P, specified as $\Phi_a^r(P)$ below

$$\Phi_a^r(P) =_{df} \mu X \bullet ((r? \to P); (a! \to X))$$

where

- r, the request signal, activates the process P, and
- a, an acknowledgement signal, indicates that P has terminated.

The process $\Phi_a^r(P)$ can be activated many times, while P cannot.

Essentially, a source program P is split into communicating processes M (P) and D (P), where

1. M (P) models the control flow of P.
2. D (P) implements variables, expressions and communication channels.

The compilation function C_a^r is simply defined by

$$C_a^r(P) =_{df} \Phi_a^r(\mathcal{M}(P))\|\mathcal{D}(P)$$

Our main goal is to prove that the target program produced by the compilation scheme serves as a refinement of the reusable version of P:

$$\Phi_a^r(P) \sqsubseteq C_a^r(P) (**)$$

The verification task is broken into three steps

- First, we prove that the parallel composition of M (P) and D (P) is a legal replacement of the source program P.

$$P \sqsubseteq (\updownarrow\mathcal{M}(P)\|\mathcal{D}(P))$$

- Then we show:

$$\Phi_s^r(P) \sqsubseteq (\Phi_s^r(P)\|\mathcal{D}(P))$$

- Finally, we are going to prove that the composite mapping $\Phi_a^r \square$M is indeed a homomorphism. Taking sequential composition as an example $\Phi_a^r \square$M is a homomorphism means that the control process of $P0;P1$ can be implemented by the parallel composition of those of its components:

$$\Phi_a^r(\mathcal{M}(P0; P1)) \sqsubseteq \Phi_{a0}^{r0}(\mathcal{M}(P0))\|SEQ\|\Phi_{a1}^{r1}(\mathcal{M}(P1))$$

where *SEQ* is designed to model the sequential composition operator:

$$SEQ =_{df} WIRE(r, r0)\|WIRE(a0, r1)\|WIRE(a1, a)$$

In this way, all the programming operators will eventually be replaced by parallel operator.

The control process M (P) involves only synchronized communications with D (P) which maintains the state of variables and channels. To avoid deadlock on the internal links between M (P) and D (P), we shall ensure that the communications on these channels satisfy the related handshake protocol. This is the topic of the next section.

7.4 Handshake protocol

In languages such as Occam, communication between processes is *synchronous* in the sense that a communication event can take place only when both the sender and the receiver agree on it. It is the synchronous nature of communication that gives these languages the expression power. In our approach, source programs are compiled into networks of primitive components, and synchronous communications of source programs are implemented by a communication protocol involving *multiple asynchronous* communication events.

Each primitive component has one or more *ports* that are used to connect it to its neighbors. Within each port, the components and their neighbors obey a handshaking protocol: events, called *request*, start the execution of the module or its neighbors, while events, called *acknowledgement*, indicate that the execution has completed. The compilation scheme ensures that no component attempts to send either a request or acknowledgement event to its partner unless the latter is waiting for such an input. With this guarantee, synchronous communications can be replaced by asynchronous ones.

The translation from synchrony to asynchrony is a necessary step in compiling our language to hardware, because the basic hardware building blocks cannot refuse input events from their environment, whereas synchronous communication implies the ability of a process to refuse to participate in a communication event. Any unexpected inputs to a circuit may lead to aberrant behavior; it is the obligation of the circuit's environment to provide input only when the circuit is waiting for it.

The simplest signal interface is the two-wire interface, which consists of one request and one acknowledgement signal. A handshake begins when a user module sends an event to a server along the wire r (for request), the user then waits a response on the wire a (for acknowledgement). When it has received an acknowledgement from the server, the handshake is complete.

7.4.1 Definition 4.1 (two wire control interface)

The handshake protocol $HP(r, a)$, used in the control interface of the target processes, behaves like a one-place buffer, repeatedly engaging in an acknowledgement event a after receipt of a request event r.

$$HP(r, a) =_{df} \mu X \bullet ((r \rightarrow a \rightarrow X) \,\|\, \mathbf{skip})$$

where the alternative **skip** is present to enable $HP(r,a)$ to respond to its parallel partners' termination request. □

A sequence of $HP(r,a)$ is still a handshake protocol.

Lemma 4.2 (sequence of handshake protocols)
$$HP(r,a);HP(r,a) = HP(r,a)$$
Proof: Algebraically using **Laws 6.4 & 6.5, 11** (twice), **1.1** and **10.1**.

Definition 4.3

A process Q satisfies the handshake protocol on (r,a) if
$$(Q\,||_{\{r,a\}} HP(r,a)) = Q$$
This fact will be denoted by $._{HP(r,a)} Q$.   □

Lemma 4.4

If $._{HP(r,a)} Q$, then
$$(Q;R)\,||_{\{r,a\}} HP(r,a) = Q;(R\,||_{\{r,a\}} HP(r,a))$$ □

Lemma 4.5

(1) $._{HP(r,a)} Q$ **skip**.

(2) $._{HP(r,a)} Q$ **stop**.

(3) If both P and Q satisfy the handshake protocol, so do $P;Q$ and $P.Q$.
$$._{HP(r,a)} (P;Q) \text{ and } ._{HP(r,a)} (P.Q).$$

(4) If Q satisfies the handshake protocol, so does the guarded recursion
$\mu X \bullet (Q;X)$.

Proofs:

Of (3):
$$(P;Q)||_{\{r,a\}} HP(r,a)$$
$$= \quad \{\textbf{Lemma 4.5} \text{ and } \vdash_{HP(r,a)} P\}$$
$$P;(Q||_{\{r,a\}} HP(r,a))$$
$$= \quad \{\vdash_{HP(r,a)} Q\}$$
$$P;Q$$

Of (4):
$$(\mu X \bullet (Q;X))||_{\{r,a\}} HP(r,a)$$
$$= \quad \{\textbf{Law 10.1}\}$$
$$(Q;\mu X \bullet (Q;X))||_{\{r,a\}} HP(r,a)$$
$$= \quad \{\textbf{Lemma 4.5} \text{ and } \vdash_{HP(r,a)} HP(r,a)\}$$
$$Q\ ((\mu X \bullet (Q;X))||_{\{r,a\}} HP(r,a))$$
$$= \quad \{\textbf{Law 11}\}$$
$$\mu X \bullet (Q;X)$$
□

In the next section, we will present the handshake protocols to pass data between the master control process M (*P*) and the data process D (*P*). For example, the handshake protocol for access of a Boolean variable has a single event *req* for requesting the current value of that variable, but contains two acknowledgement events, *val*.0 and *val*.1, for returning the value. Thus it is necessary to generalize *HP*(*r*,*a*).

Let *I* be a finite set, and let *A* be an *I*-indexed family of finite set of events, and let $B = \{(r(i), A(i))|\ i \in I\}$. The handshake protocol on *B* can then be defined as follows:

$$HP(B) \;=\; \mu X \bullet (\|_{i \in I} (r(i) \rightarrow (a : A(i) \rightarrow X)) \,\|\, \mathbf{skip}$$

This definition allows a handshake beginning with event *r*(*i*) to be completed by one of the events of *A*(*i*). A process *Q* is said to obey the handshake protocol on the set *B* if

$$Q\|_B HP(B) \;=\; Q$$

Definition 4.6 (handshake refinement)

The *handshake refinement* relation $._{HP(B)}$ between processes is defined by
$$R \sqsubseteq_{HP(B)} S \;=_{df}\; (R\|_B HP(B)) \sqsubseteq (S\|_B HP(B))$$
i.e., the process *S* is a refinement of *R* in any environment which obeys the handshake protocol *HP*(*B*). □

Lemma 4.7

$$(R\ ._{HP(B)}\ S)\ .(._{HP(B)}\ Q)\ .(R \parallel Q)\ .\ (S \parallel Q)$$

Proof:

$$R\|Q$$
$$=\quad \{\vdash_{HP(B)} Q\}$$
$$R\|(Q\|_B HP(B))$$
$$=\quad \{\mathbf{Law\ 7.5}\}$$
$$(R\|_B HP(B))\|Q$$
$$\sqsubseteq\quad \{R \sqsubseteq_{HP(B)} S\}$$
$$(S\|_B HP(B))\|Q$$
$$=\quad \{\text{similar reasoning as before}\}$$
$$S\|Q$$

□

7.5 Data processes

In order to simplify the presentation, we will only deal with the sequential subset of the language in this section, and postpone the treatment of communication and concurrency to section 8.

7.5.1   Variables

A Boolean program variable x is realized by a communication process $V(x)$, which has two handshake ports, one for reading and one for writing. To read the value of variable x, a reader send a request on $x.req$ and the variable responds wither either $x.val0$ or $x.val1$, depending upon the stored value. To response a read request correctly, the process $V(x)$ has to comply with the following requirement

(Req1) $(x.req! \rightarrow x.val?v \rightarrow Q) \parallel V(x) = (v := x); (Q \parallel V(x))$

To write a value, a writer sends a request on either $x.write.0$ or $x.write1$. After the stored value has been updated, the variable responds on $x.ack$. Accordingly, $V(P)$ needs to meet the second requirement

(Req2) $(x.write!v \rightarrow x.ack? \rightarrow Q) \parallel V(x) = (x := v); (Q \parallel V(x))$

The construction of $V(x)$ starts with design of a process $CELL(x)$ acting as the state holder of variable x

$$CELL(x) =_{df} READ(x) \; [\!] \; WRITE(x) \; [\!] \; \textbf{skip}$$

where

- $READ(x)$ is a simple handshake protocol, acting as the read interface.

$$READ(x) =_{df} x.req? \rightarrow x.val!x \rightarrow CELL(x)$$

  To avoid deadlock on the newly introduced channels $x.req$ and $x.val$, it is required that the user of $READ(x)$ obeys the handshake protocol $HP(x.req,x.val)$.

- The process $WRITE(x)$ plays the role of the write interface, and is also a handshake protocol

$$WRITE(x) =_{df} x.write?x \rightarrow x.ack! \rightarrow CELL(x)$$

  The user of $WRITE(x)$ is required to obey the handshake protocol on the channels $(x.write,x.ack)$.

$CELL(x)$ provides single user the desirable service of variable x.

Lemma 5.1   (Read and write)

Let $\textbf{Chan}(CELL(x)) \subseteq \textbf{Chan}(Q)$. Then

(1) $(x.req! \rightarrow x.val?v \rightarrow Q) \parallel V(x) = (v := x); (Q \parallel V(x))$
(2) $(x.write!v \rightarrow x.ack? \rightarrow Q) \parallel V(x) = (x := v); (Q \parallel V(x))$

Proof: Direct from the expansion law, **Law 9.2.**   □

One difficulty with building a process to implement a program variable is that it must support multiple readers and writers. We accomplish this by introducing multiplex processes *RMUL* and *WMUL*. In order to avoid interference among multiple-user request, we shall treat the read and write actions as atomic ones. For this purpose *RMUL* and *WMUL* are constructed as follows

$$RMUL(x.req, x.val) =_{df} \text{var } w \bullet (\text{skip} [\![$$

$$[\![_{i \in I}(x.req_i? \to x.req! \to x.val?w \to x.val_i!w \to RMUL))$$

$$WMUL(x.write, x, ack)) =_{df} \text{var } w \bullet (\text{skip} [\![$$

$$[\![_{j \in J}(x.write_j?w \to x.write!w \to x.ack? \to x.ack_j! \to WMUL))$$

where the sets I and J are both finite.

Putting the three processes in parallel, we finish the design of $\mathcal{V}$ (x)

$$\mathcal{V}(x) =_{df} CELL(x) \| RMUL \| WMUL$$

which is equipped with the following channels:

$$\mathbf{In}(\mathcal{V}(x)) = \{x.req_i \mid i \in I\} \cup \{x.write_j \mid j \in J\}$$

$$\mathbf{Out}(\mathcal{V}(x)) = \{x.val_i \mid i \in I\} \cup \{x.ack_j \mid j \in J\}$$

Every user of the program variable x will be allocated a pair

$$(x.req_i, x.val_i)$$

of channels for its read operation, and it is asked to obey the handshake protocol over the corresponding channels. The writers of x will be treated in a similar way.

Suppose that $\mathbf{Var}(P) = \{x_1, x_2, \dots, x_n\}$. Then the following process

$$\mathcal{V}(\mathbf{Var}(P)) =_{df} \mathcal{V}(x_1) \| \mathcal{V}(x_2) \| \dots \| \mathcal{V}(x_n)$$

is included in D (P) to represent the variable state of P.

7.5.2 Expressions

The Boolean constants *true* and *false* are realized by the *TRUE* and *FALSE* modules. The modules use the read interface and always return a particular acknowledgement – *val0* for *false* and *val1* for *true*.

$$TRUE =_{df} (true?req \to true.val1 \to TRUE) [\![\text{skip}$$

$$FALSE =_{df} (true?req \to true.val0 \to FALSE) [\![\text{skip}$$

For each Boolean expression b of P, the data process D (P) contains a component process E (b) to model its behavior. The process E (b) operates in a similar way as the read port process *READ*. For example, the process *OR* (defined below to implement $x \vee y$) first waits for a request signal from its user, and then reads the value of x from $V(x)$ to its local variable w. If w has the value true, then this value is passed to its user, otherwise it reads the value of y from V (y) and then passes it to the user.

$OR =_{df}$ **var** $w \bullet$ (**skip** $\|$

$\qquad (req? \rightarrow x.req_i! \rightarrow x.val_i?w \rightarrow$

$\qquad ((val!w \rightarrow OR) \triangleleft w \triangleright (y.req_j! \rightarrow y.val_j?w \rightarrow val!w \rightarrow OR))))$

Taking the multiple-user issue into account, we end with the following design

$$\mathcal{E}(x \vee y) =_{df} OR \| RMUL(req, val)$$

where $RMUL(req, val)$ is the process $RMUL(x.req, x.val)$ after proper channel renaming.

The module *AND*, used to implements the Boolean expression $x \wedge y$, is defined by

$AND =_{df}$ **var** $w \bullet$ (**skip** $\|$

$\qquad (req? \rightarrow x.req_i \rightarrow x.val_i?w \rightarrow$

$\qquad ((val!w \rightarrow AND) \triangleleft \neg w \triangleright (y.req! \rightarrow y.val_j?w \rightarrow val!w \rightarrow AND))))$

The module *NEG*, used to realize the negation $\neg x$, is defined by

$NEG =_{df}$ **var** $w \bullet$ (**skip** $\|$

$\qquad (req? \rightarrow x.req! \rightarrow x.val?w \rightarrow val!(1 - w) \rightarrow NEG))$

A composite expression $b = b1 \vee b2$ can be implemented in the same way as $x \vee y$ except that the former will communicate with the expression E $(b1)$ and E $(b2)$ rather than the variable processes V (x) and V (y). To avoid the name clash among the expression processes, we will rename the channels *req* and *val* in the process E (b) by *b.req* and *b.val* respectively. The definition of modules for $b1 \wedge b2$ and $\neg b$ are similar.

Let $\mathbf{Exp}(P) = \{b_1, ..., b_m\}$, we define

$$\mathcal{E}(\mathbf{Exp}(P)) =_{df} \mathcal{E}(b_1) \| \mathcal{E}(b_2) \| ... \| \mathcal{E}(b_m)$$

For a sequential program P without communication, its data process D (P) is formed by

$$\mathcal{V}(\mathbf{Var}(P)) \| \mathcal{E}(\mathbf{Exp}(P))$$

The following theorems validate our design.

Lemma 5.2   (Evaluation of expression)

Let $\mathbf{Chan}$ (D $(P)) \subseteq \mathbf{Chan}$ (Q). If $i \in RI\,(b)$, then

$$(b.req_i! \rightarrow b.val_i?v \rightarrow Q) \| \mathcal{D}(P) \; = \; (v := b); (Q \| \mathcal{D}(P))$$

Proof: From the expansion law, **Law 9.2**.

To be able to execute the expression processes in parallel with the master control process, we must make sure that the processes representing expressions have disjoint sets of channels. In particular, since most expression processes may need to access the variable processes, the allocation of the channels $x.req_i$

and $x.val_i$ turns out to be an important issue in the hardware compilation scheme. For simplicity, we assume that there are index functions RI and WI. For each variable process V (x), in addition to the channels used by expression processes the following set of channels

$$\{x.req_i, x.val_i \mid i \in RI(x)\} \cup \{x.write_j, x.ack_j \mid j \in WI(x)\}$$

is available at the disposal of the control process M (P). For an expression process $E(b)$, we take a similar convention that all channels in the set

$$\{b.req_i, b.val_i \mid i \in RI(b)\}$$

can be employed to access the current value of b. As a result, the set of free handshake channels of the data process D (P) is

$$B =_{df} \{(x.req_i, x.val_i \mid i \in RI(x) \land x \in \mathbf{Var}(P)\} \cup$$
$$\{x.write_j, x.ack_j \mid j \in WI(x) \land x \in \mathbf{Var}(P)\} \cup$$
$$\{b.req_i, b.val_i \mid i \in RI(b) \land b \in \mathbf{Exp}(P)\}$$

To simplify the proof of our compiling function in the later sections, we exploit the regularity of handshake protocols to construct SD (P), a sequential version of D (P), which does not contain parallel composition.

$$SD(P) =_{df} HP(B)$$
$$= [\![_{i \in RI(x) \land x \in \mathbf{Var}(P)} (x.req_i \rightarrow x.val_i!x \rightarrow SD(P))$$
$$[\![_{j \in WI(x) \land x \in \mathbf{Var}(x)} (x.write_j?x \rightarrow x.ack_i! \rightarrow SD(P))$$
$$[\![_{k \in RI(b) \land b \in \mathbf{Exp}(P)} (b.req_k? \rightarrow b.val_i!b \rightarrow SD(P))$$
$$[\![\mathbf{skip}$$

Since all users of D (P) must obey the handshake protocol $HP(B)$, from Lemma 4.3 we can replace D (P) by SD (P) within such a context.

Lemma 5.3 D (P) $=_{HP(B)}$ SD (P) $\square$

Lemma 5.4 SD (P); SD (P) $=$ SD (P) $\square$

7.6 Control processes

Construction of control processes is based on a translator M whose task is to replace each evaluation of a Boolean expression b by a sequence of asynchronous communications with the process E (b), and to replace every assignment to a program variable x by interactions with the variable process V (x).

Control processes of the primitive commands have straightforward definitions:

$$\mathcal{M}(\text{skip}) \quad =_{df} \quad \text{skip}$$

$$\mathcal{M}(x := b) \quad =_{df} \quad \sqcap_{i \in RI(b), j \in WI(x)} \text{ var } v \bullet$$

$$(b.req_i!.val_i?v.write_j!v \to x.ack_j? \to \text{skip})$$

The definition of M $(x := b)$ suggests that the choice of channels used to communicate with V (x) and E (b) are irrelevant. In particular, it allows the use of any available read interface to evaluate b and then write the result to x. This non-determinism allows us later to allocate a specific pair (i, j) of channel indices for implementation of M $(x := b)$.

A control process of a sequential composition is formed by those of its components:

$$\mathcal{M}(P; Q) =_{df} \mathcal{M}(P); \mathcal{M}(Q)$$

The control process M (**if** b **fi**) evaluates the Boolean guard b by interacting with the expression process E (b)

$$\mathcal{M}(\text{if } b \to P \text{ fi}) \quad =_{df}$$

$$\sqcap_{i \in RI(b)} \text{ var } w \bullet (b.req_i! \to b.val_i?w \to (\mathcal{M}(P) \lhd w \rhd \text{chaos}))$$

Conditional with multiple branches behaves as if its Boolean guards are calculated sequentially.

$$\mathcal{M}(\text{if } b \to P \| BG \text{ fi}) =_{df}$$

$$\sqcap_{i \in RI(b)} \text{ var } w \bullet (b.req_i! \to b.val_i?w \to (\mathcal{M}(P) \lhd b \rhd \mathcal{M}(\text{if } BG \text{ fi})))$$

The process M(**do** BG **od**) evaluates its guards in sequel and terminates when none of them is true. As expected, it is implemented by a tail recursion

$$\mathcal{M}(\text{do } BG \text{ od}) =_{df} \text{ var } w \bullet \mu X \bullet \mathcal{M}(BG)$$

where

$$\mathcal{M}(b \to P) =_{df}$$

$$\sqcap_{i \in RI(b)} (b.req_i! \to b.val_i?w \to ((\mathcal{M}(P); X) \lhd w \rhd \text{skip}))$$

$$\mathcal{M}((b \to P) \| BG) =_{df}$$

$$\sqcap_{i \in RI(b)} (b.req_i! \to b.val_i?w \to ((\mathcal{M}(P); X) \lhd w \rhd \mathcal{M}(BG)))$$

From the definition of M and Lemma 4.5 we conclude that all the control processes M (P) obey the handshaking protocol $HP(B)$, where the set B was defined in the previous section.

Lemma 6.1

$$(\text{M}(P); Q) \| \text{SD}(P) = (\text{M}(P) \| \text{SD}(P)); (Q \| \text{SD}(P)) \qquad \square$$

Our compilation strategy is validated by the following theorem.

Theorem 6.2 (Correctness of the control processes)

$$P .\text{M}(P) \| \text{D}(P)$$

Outline of proof: The proof is based on structural induction using the following base case and inductive steps.

(1) Basic case: $P = (x := b)$

(2) Inductive Step:

(2.1) $P = Q ; R$

(2.2) $P = \textbf{if } (b \rightarrow Q) \textbf{ fi}$

(2.3) $P = \textbf{if } (b \rightarrow Q \parallel BG) \textbf{ fi}$

(2.4) $P = \textbf{do } (b \rightarrow Q) \textbf{ od}$

(2.5) $P = \textbf{do } (b \rightarrow Q \parallel BG) \textbf{ od}$ □

Now we can establish the correctness of the compiling function M.

Theorem 6.3 (correctness of compiling function)

$$\Phi^r_a(P) \sqsubseteq \Phi^r_a(\mathcal{M}(P)) \| \mathcal{D}(P)$$

Proof: Using **Lemmas 4.7 & 5.3**, **Law 9.1**, **Lemma 6.1**, **Lemmas 4.7 & 5.3** (again), **Theorem 6.2** and **Law 9.2**.

The process $\Phi^r_a(P)$ enjoys a number of algebraic laws. The first one says the additional wires used to connect $\Phi^r_a(P)$ with its environment has no effect.

Lemma 6.4

Let $r0$ and $r0$ be fresh events. Then

$$\Phi^r_a(P) =_{HP(r,a)} WIRE(r, r0) \| \Phi^{r0}_{a0}(P) \| WIRE(a0, a)$$

The second law enables us to implement the sequential composition by the parallel composition.

Lemma 6.5

Let Q and R be processes with disjoint sets of variables and channels. If h is a fresh event, then

$$\Phi^r_a(Q; R) =_{HP(r,a)} \Phi^r_h(Q) \| \Phi^h_a(R)$$

7.7 Hardware device

In this section we are going to divide the control process $\Phi^r_a(\text{M}(P))$ into a set of primitive handshake modules within an environment obeying both $HP(r,a)$ and $HP(B)$, where B denotes the set of channels used to access the data process D(P), and was defined in Section 4. We define

$$(Q \sqsubseteq_{Env} P) \quad \textbf{iff} \quad (Q \|_S PROT) \sqsubseteq (P \|_S PROT)$$

where $S =_{df} B \cup \{r, a\}$

and $PROT =_{df} HP(B) \parallel HP(r,a)$ describes the behavior of the environment which obeys the handshake protocols $HP(B)$ and $HP(r,a)$.

Lemma 7.1

If Q obeys the handshake protocols $HP(B)$ and $HP(r,a)$, then

$$(Q \sqsubseteq_{Env} P \quad \textbf{iff} \quad Q \sqsubseteq (P \parallel_S PROT)$$

The main objectives of our design are to preserve the modular structure of the source program, and to generate a small number of hardware devices. We have already presented a number of handshake modules such as $CELL(x)$ and OR in Section 4. The following includes four further examples.

First, the command **skip** can be implemented by a wire.

Theorem 7.2 (skip)

$$\Phi_a^r(\mathcal{M}(\textbf{skip})) =_{Env} WIRE(r, a)$$

Proof:

$$WIRE(r, a) \parallel_{\{r, a\}} HP(r, a)$$
$$= \quad \{\textbf{Law 8.1}\}$$
$$r \rightarrow a \rightarrow (WIRE(r, a) \parallel_{\{r, a\}} HP(r, a))$$
$$= \quad \{\textbf{Law 11}\}$$
$$\mu X \bullet (r \rightarrow a \rightarrow X)$$
$$= \quad \{\textbf{Law 8.1}\}$$
$$\mu X \bullet (r \rightarrow a \rightarrow X) \parallel_{\{r, a\}} HP(r, a)$$
$$= \quad \{\text{Def of } \mathcal{M}(\textbf{skip})\}$$
$$\Phi_a^r(\mathcal{M}(\textbf{skip})) \parallel_{\{r, a\}} HP(r, a)$$

$\square$

The control process of $x := b$ can be implemented by a set of wires, where a pair **Error!** of channels is chosen as the read interface with E (b), and **Error!** as the write interface with V (x).

Theorem 7.3 (assignment)

$$\Phi_a^r(\mathcal{M}(v := b)) \sqsubseteq_{Env} ASGN, \text{ where}$$
$$ASGN =_{df} WIRE(r, b.req_i) \parallel WIRE(b.val_i, x.write_j) \parallel WIRE(x.ack_j, a)$$

Proof: Similar to Theorem 7.2.

Surprisingly, the sequential composition operator can also be implemented by a set of wires.

Theorem 7.4 (sequential composition)

Let $P0$ and $P1$ be processes with disjoint channels. If h is not used by $P0$ and $P1$, then

$$\Phi_a^r(P0; P1) \sqsubseteq_{Env} \Phi_h^r(P_0) \| \Phi_a^h(P1)$$

Proof: From **Lemma 6.5**.

The final two theorems show that the control processes of conditional and iteration can be implemented by the hardware device $M(in1, in2: out)$, which merges signals received from its input ports $in0$ and $in1$

$$M(in0, in1 : out) =_{df} \mu X \bullet (in0?!) [\![(in1?! \to X)$$

Theorem 7.5 (conditional)

Let $i \in RI(b)$. If none of $\{r, b.req_i, b.val_i, a0, a1, a\}$ is used by Q and R, then

(1) $\Phi_a^r(b.req_i! \to (b.val_i.1 \to Q \| b.val_i.0 \to \mathbf{chaos}))$

$\sqsubseteq_{Env} M(r, b.val_i.0 : b.req_i) \| \Phi_a^{b.val_i.1}(Q)$

(2) $\Phi_a^r(b.req_i! \to (b.val_i.1w \to Q \| b.val_i.0 \to R))$

$\sqsubseteq_{Env} WIRE(r, b.req_i) \| \Phi_{a1}^{b.val_i.1}(Q) \| \Phi_{a0}^{b.val_i.0}(R) \| M(a0, a1 : a)$

Proof: Define

$$NEW =_{df} RHS \|_S PROT$$

We are going to establish the following fact
$NEW.LHS$

For notational convenience, we will drop the port parameters of M and Φ in the proof below.

(1) NEW

$=$ {**Law 8**}

   $r \to (\Phi(Q) \| (b.req_i \to M)) \|_S ((a \to HP(r, a)) \| HP(B)))$

$=$ {**Law 8**}

   $r \to b.req_i! \to ((\Phi(Q) \| M) \|_S$

   $\qquad\qquad ((a \to HP(r, a)) \| ((b.val_i.0 \to HP(B)) [\![(b.val_i.1 \to HP(B)))))$

$=$ {**Laws 8 and 9**}

   $r \to b.req_i \to$

   $(b.val_i.0 \to ((\Phi(Q) \| (b.req_i \to M)) \|_S ((a \to HP(r, a) \| HP(B)))) [\![$

   $b.val_i.1 \to Q; (a \to NEW))$

$\sqsupseteq$ {**Laws 2 and 3.5**}

   $r \to b.req_i \to (b.val_i.1 \to Q \| b.val_i.0 \to \mathbf{chaos}); (a \to NEW))$

from which together with **Law 10.2** it follows that

$$NEW$$

$$\sqsupseteq \quad \mu X \bullet (r \to b.req_i \to (b.val_i.1 \to Q \,[\![\, b.val_i.0 \to \mathbf{chaos})); (a \to X))$$

$$= \quad \{\text{Def. of } \Phi_a^r\}$$

$$\Phi_a^r(b.req_i! \to (b.val_i.1 \to Q \,[\![\, b.val_i.0 \to \mathbf{chaos}))$$

(2) Similar to (1).

Theorem 7.6 (iteration)

Let $i \in RI(b)$. If none of $\{r, \hat{a}, a, b.req_i, b.val_i\}$ is used by Q, then

$$\Phi_a^r(\mu X \bullet \ (b.req_i! \to (b.val_i.1 \to (Q; X) \,[\![\, b.val_i.0 \to \mathbf{skip})$$

$$\sqsubseteq_{Env} \ M(r, \hat{a} \ b.req_i) \| \Phi_{\hat{a}}^{b.val_i.1}(Q) \| WIRE(b.val_i.0, a).$$

Proof: Define

$$U \ =_{df} \ \mu X \bullet (b.req_i \to (b.val_i.1 \to (Q; X) \,[\![\, b.val_i.0 \to \mathbf{skip}))$$

$$V1 \ =_{df} \ \Phi_a^r(U)$$

$$W1 \ =_{df} \ U; (a1))$$

$$V2 \ =_{df} \ (M(r, \hat{a} \ b.req_i) \| \Phi_{\hat{a}}^{b.val_i.1}(Q) \| WIRE(b.val_i.0, a)) \,\|_S PROT$$

$$W2 \ =_{df} \ (b.req_i(r, \hat{a} \ \_req_i) \| \Phi_{\hat{a}}^{b.val_i.1}(Q) \| WIRE(b.val_i.0, a))$$

$$\|_S((a(r, a)) \| HP(B))$$

We are going to show that $(V1, W1)$ and $(V2, W2)$ satisfy the same guarded recursive equation.

$$V1$$

$$= \quad \{\text{Def of } \Phi \text{ and } \mathbf{Law\ 10.1}\}$$

$$(r \to U); (a \to V1)$$

$$= \quad \{\mathbf{Law\ 6.7}\}$$

$$r \to W1$$

$$W1$$

$$= \quad \{\mathbf{Law\ 10.1}\}$$

$$(b.req_i \to (b.val_i.1 \to (Q; U) [\![b.val_i.0 \to \mathbf{skip}));$$

$$(a \to V1)$$

$$= \quad \{\mathbf{Laws\ 3.4 \text{ and } 6.7}\}$$

$$b.req_i \to (b.val.1 \to (Q; W1) [\![b.val_i.0 \to (a \to V1))$$

From **Laws 8** and **9** it follows that:

$$V2 \;=\; r \to b.req_i \to$$

$$((b.val_i.1 \to (Q; V2)) \, \| \, (b.val_i.0 \to a \to U1))$$

$$W2 \;=\; b.req_i \to ((b.val_i.1 \to (Q; V2)) \, \| \, (b.val_i.0 \to a \to U2))$$

From the unique fixed point theorem **Law 11** we obtain $V1 = V2$ as required.

7.8 Conclusion

Hardware compilation is an exciting development in the array of techniques available to generate a implementation from a high-level description. It allows hardware to be generated very quickly from software. What is more, it is possible to formally prove the relationship between a high-level program (software) and a low-level *netlist* of components connected with wires (hardware) is correct [3,5].

In this chapter we have presented a small programming language that can be compiled into hardware. A set of algebraic laws are given for the language that allow programs to be transformed in a provably correct manner. A compilation strategy from this language to a hardware description in the form of a netlist is given for each of the constructs in the language. These are posited as theorems that can be proved correct in an algebraic style. Some sample proofs are given.

For the future, it is expected that hardware will increasingly be generated from high-level descriptions, especially when the available design time is limited, costs need to be kept down and speed of execution is not an overriding factor. In addition, this approach can help to raise the level of confidence in the correctness of the implementation since, as demonstrated in this chapter, it is possible to mathematically prove that the transformation from the high-level description to the low-level implementation is correct, not just for individual cases, but for all designs that are generated following such a compilation scheme.

7.9 References

[1]   K. van Berkel, J. Kessels, M. Roncken, R. W. J. J. Saeijs and F. Schalij. The VLSI language Tangram and its translation into handshake circuits. In *Proceedings of the European Design Automation Conference*, (1991).

[2]   K. van Berkel. Handshake Circuits: An Asynchronous Architecture for VLSI Programming. Cambridge University Press, (1993).

[3]   J. P. Bowen and J. He. An approach to the specification and verification of a hardware compilation scheme. *The Journal of Supercomputing* 19(1):23–39, (2001).

[4]   S. D. Brookes, C. A. R. Hoare and A. W. Roscoe. A theory of communicating sequential processes. *Journal of the ACM* 31(3):560–599, (1984).

[5]   J. He. An algebraic approach to the VERILOG programming. In *Formal Methods at the Crossroads*, Lecture Notes in Computer Science 2757, pages 65–80, Springer-Verlag, (2003).

[6]   C. A. R. Hoare. *Communicating Sequential Processes.* Prentice Hall International Series in Computer Science, (1985).

[7]   Inmos Limited. *Occam 2 Reference Manual.* Prentice Hall International, (1988).

[8]   Intel Corporation. *1994 Annual Report.* (1995).

[9]   A. J. Martin. Programming in VLSI: From communicating processes to delay-insensitive circuits. In *Developments in Concurrency and Communication*, C. A. R. Hoare (ed.), pages 1–64, Addison-Wesley, (1990).

[10]   I. Page and W. Luk. Compiling occam into field-programmable gate arrays. In *Field-Programmable Gate Arrays*, W. Moore and W. Luk (eds.) pages 271–283, Abingdon EECS Books, (1991).

[11]   S. Palnitkar. Verilog HDL, 2nd edition. Prectice Hall PTR, (2003).

[12]   V. A. Pedroni, *Circuit Design with VHDL*. The MIT Press, (2004).

[13]   A. W. Roscoe and C. A. R. Hoare. Laws of Occam programming. *Theoretical Computer Science* 60:177–229, (1988).

[14]   A. W. Roscoe. *The Theory and Practice of Concurrency.* Prentice Hall International Series in Computer Science, (1997).

[15]   A. Tarski. A lattice-theoretical fixpoint theorem and its applications. *Pacific Journal of Mathematics*, 5:285–309, (1955).

[16]   S. W. Weber, B. Bloom and G. Brown. Compiling Joy into silicon. In *Advanced Research in VLSI and Parallel Systems*, T. Knight and J. Savage (eds.), MIT Press, (1992).

[17]   Xilinx Inc. *Programmable Logic Devices, FPGA & CPLD.* www.xilinx.com, San Jose, California, USA, (2005).

8 Formal Methods for UML

Authors
Mª Encarnación Beato
Escuela Universitaria de Informática
Universidad Pontificia de Salamanca, Spain

Manuel Barrio-Solórzano, Carlos E. Cuesta, and Pablo de la Fuente
Departamento de Informática
Universidad de Valladolid, Spain

Summary
The use of the UML specification language is very widespread due to some of its features. However, the ever more complex systems of today require modelling methods that allow errors to be detected in the initial phases of development. The use of formal methods makes such error detection possible but the learning cost is high.

This paper presents a tool, which avoids this learning cost, enabling the active behavior of a system expressed in UML to be verified in a completely automatic way by means of formal method techniques. It incorporates an assistant for the verification that acts as a user guide for writing properties so that she/he needs no knowledge of either temporal logic or the form of the specification obtained.

Keywords: formal methods, formal specifications, Formal Verification, Model Checking, SMV.

8.1 Introduction

Unified modeling language (UML) is a widely used graphic language accepted as the standard for modeling any kind of software system, following an object-oriented philosophy. It is organized in diagrams that provide different views covering all aspects of a development.

Hossam A. Gabbar (ed.), Modern Formal Methods and Applications, 177–196.

UML has unquestionable advantages as a visual modeling technique, and this has meant that its applications have multiplied rapidly since its inception. To the characteristics of UML itself must be added numerous tools that exist in the market to help in its use (Rational Rose, Argo UML, Rhapsody ...). However, unfortunately, none of them guarantee specification correction.

It is widely accepted that error detection in the early phases of development substantially reduces cost and development time, as the errors detected are not transmitted to or amplified in later phases. It would thus be very useful to have a tool that would allow the integration of this semi-formal development method with a formal method to enable system verification. This paper presents a tool to carry out this integration by providing a formal framework in which to verify the UML active behavior.

The formal specification language chosen is SMV as it has the adequate characteristics for representing the active behavior of a specification in UML. The main reason for this is that it is based on labeled transition systems and because it allows the user's own defined data types to be used, thus facilitating the definition of variables. It also uses symbolic model checking for the verification, which means that the test is automatic, always obtains an answer and (more importantly, should the property not be satisfied) generates a means of identifying the originating error.

The tool carries out, with no intervention on the user's part, a complete, automatic transformation of the active behavior specified in UML into an SMV specification, focusing mainly on the reactive systems in which the active behavior of the classes is represented through state diagrams, while activity diagrams are used to reflect the behavior of class operations. XMI (XML Metadata Interchange) is used as the input format, thus making it independent of the tool used for the system specification. Nowadays, most UML CASE tools on the market incorporate the possibility of exporting the specification through XMI (Rational Rose, Argo UML, Rhapsody ...).

On the other hand, the tool has a versatile assistant that guides the user in writing properties to be verified using temporal logic. The verification is carried out in such a way that the user needs no knowledge of either formal languages or temporal logic to be able to take advantage of its potential; something which has traditionally been difficult to overcome when deciding on the use of formal methods. In addition, notions of the form of the specification obtained are unnecessary: that is, knowledge of the internal structure of variables or modules obtained is not required for verification. The tool's architecture can be seen in Figure 8.1.

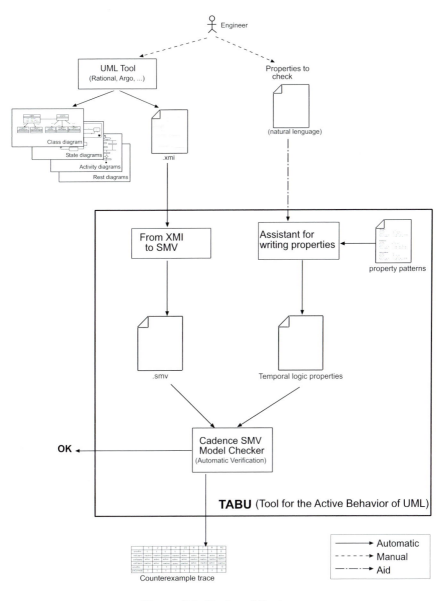

Figure 8-1. Tool architecture

The rest of the paper is structured as follows. Section two presents the philosophy used to carry out the transformation of the UML active behavior to SMV, with the help of simple examples to aid understanding. Section three focuses on verification, showing how the assistant functions and the classification of properties. Finally, the conclusions are presented along with possible future work.

8.2 From UML to SMV

Three kinds of diagram are taken into account when transforming the active behaviour from UML to SMV: class, state and activity diagrams. The first provides information concerning the elements that make up the system and their relationships while the second and third provide information about the behaviour, through time, of each of those elements.

In order to show the functionality of the tool TABU, a banking ATM example has been developed. For the system to work correctly, the user is supposed to insert a card followed by the pin number. If this number is correct, the user will be allowed to push the button `getMoney`. At this point the system checks the card balance and ATM money availability, updates them and delivers both the money and the card.

If the user introduces an erroneous pin number, the system will ask for the correct number once more. If a wrong pin number is introduced three times the function is aborted and the card kept. Additionally, users can always push the button `cancel` producing an error signal and forcing the system to deliver the card back to the user. Figures 8.2, 8.3, 8.4, 8.5 and 8.6 show the UML diagrams that specify this behaviour.

8.2.1 Classes

The fundamental concept taken as our starting point is that of the active class, that is, where objects have one or more processes or threads of execution and can thus initiate control activities. The behaviour of each active class is reflected in a different SMV module, which in turn is instantiated in the main module by each of the class objects. Although it is possible to change the number of objects in a dynamic way in UML, this is not possible in SMV, as it cannot support such dynamism. This means that the number of instances of the classes must be fixed and known a priori.

Each SMV module, representing a class, needs the signals the class receives as its input parameters, and those the class emits as output parameters. Thus, the said signals are reflected in the class diagram using the stereotypes <<send>> and <<signal>> as shown in figure 8.2. Here, the signal updateBalanceCard correspond to the signal emitted by the Card class, while checkCard y pushCancel are the received.

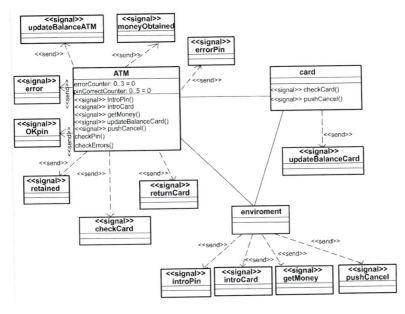

Figure 8-2. Class Diagram

An additional class called environment also has to be included. It has no associated behaviour and contains details of the signals produced outside the system and which are input signals.

Classes can have attributes, specifying the type of data and the initial value, if they have one. These attributes modify their value in line with the system's evolution, typically through the firing of a transition. Their evolution will thus be reflected in a similar way to that of the state machines discussed below.

On the other hand, a class may include operations whose behaviour will be reflected in an activity diagram. This is activated by the occurrence of an activity carried out by a state of the machine of that class. The activities are dealt with in section 1.2.7.

8.2.2 State machines

State machines represent the active behaviour through time of the objects in a class. They are principally the states through which they pass in response to events or signals. Thus, to correctly control the evolution of a state machine, the state it is in at any given moment must be known. This is achieved by using a separate variable to store this information for each machine.

In addition, the fact that combined states, both sequential and concurrent, may appear within a machine must be taken into account. This means it is also necessary to deal with the submachines of the said states. These will be dealt with following the same reasoning as for the main machine, with the exception of the peculiarities they possess with respect to activation and deactivation.

A state machine is initiated using the SMV operator `init` with the variable that controls the machine's states. If the machine is not initially active, or it has no initial state, then it will be initiated in an indeterminate state we shall call `dontKnow`. Thus, the first thing to determine for each machine is whether or not it is initially active. The most external machine, that describes the class behaviour, is always active, while the rest are initially active if their ancestors were initially active.

As for their evolution, the operator `next` is used. This represents the value taken by the variable in the following step. The syntaxes used is as shown below:

```
next(est_<machine>):= case {
    t₁ | t₁₁ | ... | t₁ₙ : <state₁>;
    :
    :
    tₙ | tₙ₁ | ... | tₙₙ : <stateₙ>;
    default: est_<machine>;
};
```

Where t_1, t_{11}, ... t_{nn} represent the firing of transitions t_1, t_{11}, ... t_{nn}. The first transitions, up to t1n correspond to those entering the state <state$_1$> and t_n ... t_{nn} to those entering the state <state$_n$>. The evolution of a machine's states through time is thus represented. All that is needed is to look at the different transitions entering each state.

The order `default` represents the behaviour where there is no change of state, that is, when no transition present in the machine is fired and it remains in the same state during the following step.

For example, the representation in SMV of the external state machine of ATM class, (see Figure 8.3) is as follows:

```
init(st_ATM)  := inactive;
next(st_ATM)  := case {
    tr_G_9  : checkingBalance;
    tr_G_7  : waitingOperation;
    tr_G_29 : inactive;
    tr_G_16 : inactive;
    tr_G_2  : active;
    tr_G_12 : returningCard;
    tr_G_14 : returningCard;
    default : st_ATM;
};
```

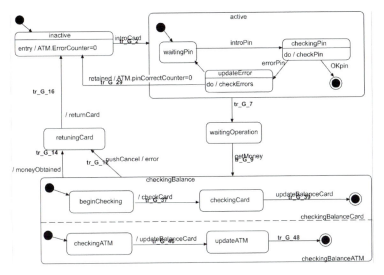

Figure 8-3. Statemachine ATM class

In fact, this is the scheme followed by the most external state machine. It is also important to take into account the conditions that cause each submachine to be activated or deactivated.

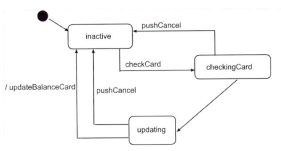

Figure 8-4. Statemachine Card class

8.2.3 Activation of a submachine

There are two different ways to activate a state submachine:
- Explicit activation. This is when the submachine is activated concretely in one of its states. This happens when a transition is fired which has that state as its destination and in the step prior to the firing the submachine was inactive.
- Activation by default. Here the submachine is activated, passing to its initial state if it has one. Otherwise it passes to an indeterminate state.

Explicit activation is already contemplated in the previous scheme, as it occurs when a transition is fired whose destination is a state of the submachine. The case, which is not reflected is that of activation by default. There are numerous actions, which cause activation of a submachine by default:

1. The firing of input transitions to the state containing the submachine.
2. If the state containing the submachine is initial, the input transitions to its father or ancestors, as long as these are still initial.
3. If the containing state is initial and has concurrent antecedents, and all the antecedents up to the concurrent one are initial, the firing of transitions that explicitly activate any of the sibling regions of that concurrent antecedent.
4. If the containing state is concurrent, the firing of transitions that activate any of the concurrent sibling regions.
5. Finally, it can be activated by its history. For instance, if the containing state has a sibling which is a history state, the input transitions to this history state could activate it.

8.2.4 Submachine deactivation

As for deactivation, the general reasons for deactivation of a submachine are:

1. The firing of any of the output transitions of any of the ancestors.
2. The firing of the transitions that originate in any of the descendants and do not have as their destination a state, which is a descendant of that submachine.
3. The firing of transitions that deactivate it by concurrence, that is, they deactivate any of the sibling regions, even though they are not directly related.

For instance, the behaviour of the concurrent submachines, checkingBalanceCard and checkingBalanceATM (see Figure 8.3) is represented as follows:

```
/***** Evolution of statemachine for state:
chechingBalanceATM *****/
   init(st_chechingBalanceATM)  := DontKnow;
   next(st_checkingBalanceATM)  := case {
      tr_G_12 : DontKnow;
      tr_G_14 : DontKnow;
      tr_G_9 : checkingATM;
      tr_G_46 : updateATM;
      tr_G_48 : FINAL;
      default : st_checkingBalanceATM;
   };
```

```
/***** Evolution of statemachine for state:
chechingBalanceCard*****/
   init(st_chechingBalanceCard) := DontKnow;
   next(st_chechingBalanceCard) := case {
      tr_G_12 : DontKnow;
      tr_G_14 : DontKnow;
      tr_G_9 : beginChecking;
      tr_G_37 : checkingCard;
      tr_G_39 : FINAL;
      default : st_chechingBalanceCard;
   };
```

Where the transition that activates the submachines by default is tr_G_9 and transitions tr_G_12 and tr_G_14 deactivate them.

8.2.5 Transitions

Transitions are used to connect machine states. Their firing causes the machine to pass from the state of origin to the state of destination. It also causes the effects associated with it. For a transition to be fired the source state must be active and, if there is a trigger event or a guard condition, the trigger event must occur. At that moment, the guard condition must also be met.

The guard condition is an expression evaluated from a Boolean result. Given that in UML no formal grammar is defined for such expressions, it will be assumed that they are written following SMV grammar and, if the value of an attribute is accessed, then the nomenclature is <class>.<attribute>.

On the other hand, three different conditions that fire a transition can be distinguished. For all of them, the firing is represented in SMV by the fact that a variable, which identifies the transition, takes a certain value.

- Fire by means of a firing event. This is the most general case where the transition has an event, which causes the firing.
- Fire due to termination. If the transition has no firing event, it can be fired when its state of origin ends the activity it was carrying out, or when the internal submachine of the state of origin of the transition ends.
- Fire by the passage of time. These are known as after transitions and are fired when a particular period of time has passed since attaining the state of origin of the transition. To represent this passage of time, a variable is used which acts as a chronometer and which increases with each step, as long as the state of origin of the transition is still active.

For example in the statechart of the Figure 8.3, the fire transition tr_G_14 is represented as:

```
   tr_G_14:=in_checkingBalance &
in_FINALcheckignBalanceCard

                    & in_FINALcheckignBalanceATM;
```

8.2.6 Actions

The evolution of an active object can lead to different actions, including sending signals and modifying the value of class attributes. With regard to sending signals, it can happen in any of the following situations:

1. The firing of a transition, if the signal is among the transition effects
2. The activation of a state, if the signal is among its entry actions; and
3. The deactivation of a state, if the signal is among its exit actions.

Taking into account that both state activation and deactivation are due to the firing of some transition, signal evolution can be represented in a similar way to state machine evolution.

As for modifying the value of an attribute, very much the same philosophy can be followed. This means that it will be specified through the use of the SMV operators `init` and `next`. Attributes will be initialised with `init` if they have an initial value in the class diagram, whereas their evolution (`next`) will depend on the firing of transitions. For instance, the SMV behaviour for the attribute `errorCounter` in class `ATM`, which keeps track of how many wrong consecutive pin numbers have been introduced, is the following (see Figures 8.2, 8.3 and 8.5).

```
/***** Attribute: contadorErrores *****/
   init(ATM_errorCounter):=0;
   next(ATM_errorCounter) := case {
      tr_G_57: ATM_errorCounter +1;
      tr_G_29: 0;
      tr_G_16: 0;
      default : ATM_errorCounter;
   };
```

The same reasoning is followed in the evolution of the value of the attributes of the class that can be modified in the input or output actions of a state and in the effects of a transition.

8.2.7 Activity diagrams

A class operation control flow can be modelled using activity diagrams, which, fundamentally, show the control flow between activities. Its SMV specification can be found in the module that reflects class behaviour. These activity diagrams are activated whenever a call to an activity is produced within a state using the notation do<activity>.

Activity diagrams can be considered as a special case of state diagrams where the majority of states are activity states and most transitions are fired by termination. So the mechanism used to represent them is similar to that used for state machines. The only difference is that, for concurrent evolution, the special states of division and union (`fork` and `join`) are used. They are activated whenever any state, which has a call to this activity inside it is activated.

Likewise, they are deactivated whenever a transition is produced that deactivates the state in which it is contained.

The activity diagrams to the operation of the ATM class, `checkErrors` and `checkPin` are represented in the Figures 8.5 and 8.6.

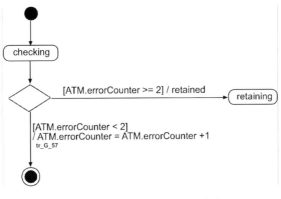

Figure 8-5. checkErrors activity

8.3 Verification

Having obtained a system specification in a formal language with a solid mathematical basis means that it is possible to check whether the system complies with certain desirable properties. As with the formal specification methods, the increasing complexity of software systems requires the development of new verification methods and tools to carry it out either automatically or semi-automatically.

In this paper, verification is carried out using the SMV tool model checker (Cadence SMV[3]). With this, it is possible to make the verification process completely automatic. That is, given a property, a positive or negative reply is always obtained.

The property must be expressed in a temporal logic present in SMV, CTL (Computation Tree Logic) or LTL (Linear Temporal Logic). This property writing is not a trivial problem. To write them correctly, advanced knowledge of logics and the type of specification obtained from the system is necessary.

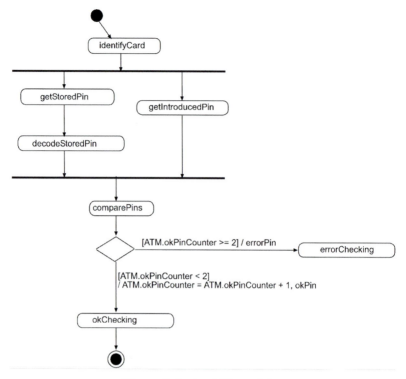

Figure 8-6. checkPin activity

Our tool overcomes this problem as it has an assistant that guides the user through the writing of properties until the property to be verified is finally obtained following the appropriate syntax.

Our starting point was the pattern classification proposed by Dwyer et al [7] to which our own cataloguing of the different properties to be automatically verified has been added. In any case, an expert user also has the possibility of introducing manually the property to be checked.

8.3.1 Property patterns

Dwyer et al [7] establish a first classification between patterns of occurrence and order. Occurrence patterns describe properties with respect to the occurrence of a state or signal during the evolution of a system. These include absence (never), universality (always), existence (sometimes) and bounded existence (appearing a certain number of times).

---

[3] http://www-cad.eecs.berkeley.edu/~kenmcmil/smv/

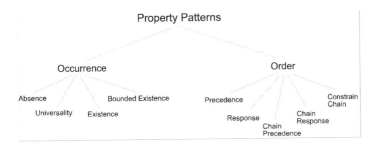

Figure 8-7. Property patterns

Order patterns establish properties with respect to the order in which they occur. They include: precedence (s precedes p), response (s responds to p), and combinations of both: chain precedence (s and t precede p or p precedes s and t), chain response (s and t respond to p or p responds to s and t), and constrain chain (s and t without z respond to p).

On the other hand, each kind of pattern has a scope of application, which indicates the system execution on which it must be verified. There are five basic scopes:

1. Global. The entire program execution.
2. Before R. The execution up to a given property R.
3. After Q. The execution after a given property Q.
4. Between Q and R. Any part of the execution from a given property Q to another given property R.
5. After Q until R. Like between but the designated part of the execution continues even if the second property does not occur.

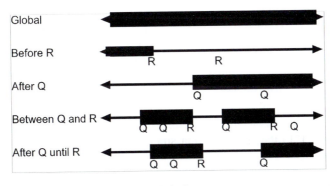

Figure 8-8. Scope

Once the type and scope of the property to be verified are known, we can complete the property pattern. As an example, Table 8.1 shows the LTL Absence pattern. The rest of patterns can be found in [3].

Table 8.1 Absence pattern

Global	`G(~P)`
Before R	`F(R) -> (~P U R)`
After Q	`G(Q -> G(~P))`
Between Q and R	`G((Q & ~R & F R) -> (~P U R))`
After Q until R	`G(Q & ~R -> (~P W R))`

8.3.2   Property classification

The different properties to be verified have been catalogued to establish limits for the scopes (Q and R) and to specify the order of properties when more than one must be determined (s, t o z), so that the user does not need to know or understand the structure of the specification obtained in SMV to carry out verification

The established property types are:

- A state machine with one active object is in a particular state. The information that has to be introduced is the name of the class to which the machine belongs, the identification of the object being referred to and the name of the state.
- An object activity is in a particular state. The information that must be introduced by the user is the same as in the above case.
- A signal or event is produced. Only the name of the signal is needed.
- Value comparison of an attribute. For this, the user must introduce the class name, the object identification, the name of the attribute, the kind of comparison (equality, inequality, less than, less than or equal to, greater than, greater than or equal to) and the value (see Figure 8.10).

The tool will automatically generate the property in the adequate format, in accordance with the chosen option and the selected pattern and scopes. Once we have the properties to be verified, it is possible, using the tool itself, to execute the SMV model checker to carry out the verification. If the property is not satisfied, it generates a trace showing a case where it is not verified.

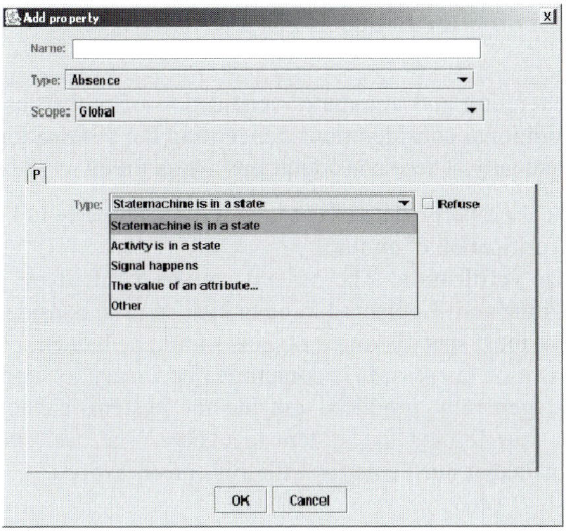

Figure 8-9. Classification of properties to be added

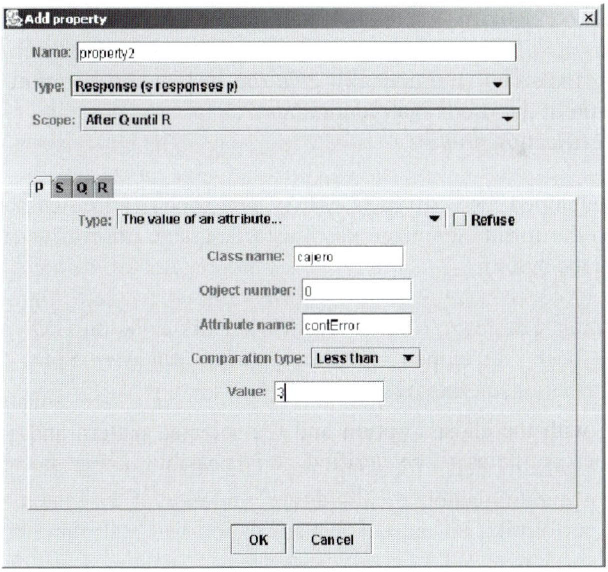

Figure 8-10. Value comparison of an attribute

8.3.3 Other considerations

Besides the generation and automatic verification of properties, the tool also allows some additional considerations concerning the verification to be carried out, also automatically. These considerations are as follows:

1. Assume a property. The user can assume a property to be true to assist in the verification of another.
2. Modular verification. The system can be verified by parts. That is, some of the active objects can be hidden so as to consider only the rest. The user must specify which objects should be hidden.
3. Treatment of fairness. It is sometimes necessary to suppose that som signals, generally produced outside the system, cannot be infinitely absent, thus forcing the system to evolve. The user should introduce the signals that must comply with this consideration.

8.4 Related work

Most works dealing with UML verification using formal method techniques focus on state diagram verification. Many such papers are based on previous works that, using formal methods, verified the classic *statechart* Harel [10] on which the UML state diagrams are based.

Yet today, as the use of UML is so widespread, researchers are looking to take a step further in UML verification with formal methods. Research is focusing on obtaining tools that automatically give the system specification in a formal language without the user's intervention and, as far as is possible, to achieve an automatic verification process.

The tools developed for verifying UML system specifications can be classified according to the formal language used, as a language prior to verification, for representing the system.

- Promela Language (tool Spin). Most work done on UML verification has been developed for the *model checker* Spin. The main contributions in this field are:

  - vUML [13,17]. vUML is a tool for the automatic verification of UML models, focusing on state diagrams. It is easy to use, generating automatically a representation of the system in Spin and performing an automatic verification to check that states of error are never reached and that states catalogued as desirable are reached.

- Latella, Majzik and Massink[12] work with UML state diagrams, encoding them through hierarchical automata (HA) from which they generate the specification for Spin. The transformation is not carried out automatically, although they have achieved this in later works using XMI [6].

- HUGO[18,11]. This project includes a set of tools to apply *model checking* to UML state diagrams and collaborating ones. The latter are used to check whether the interaction represented in the collaboration diagram can be performed using state machines. That is, for the verification. The step from UML to Spin is carried out automatically using XMI.

- SMV language. There are also some works in the literature which try to verify UML by using the SMV model checker.

  - VeriUML [5] are a set of integrated tools developed in the University of Michigan that allow UML state diagrams to be verified and to check whether the model is syntactically correct.

    This first set of tools, developed in the year 2000, was later extended [19,20], allowing verification of the model's static part. The points of entry for this approach are the class diagram, the restrictions imposed on the said diagram written in OCL and an object diagram, all in XMI format.

  - TCM (*Toolkit for Conceptual Modelling*), is a set of tools developed by R. Eshuis [8] which allows activity diagrams to be verified by converting them to transition systems that can be verified using nuSMV. The transformation is automatic, although it is not based on XMI.

If a comparison is made between the work presented in this paper and the above-mentioned work, it can be concluded that the main characteristics of this paper focus on the possibility of performing an automatic verification of the behaviour of a UML specification in which the said behaviour is reflected through state and activity diagrams and is also semi-transparent for the user. Most works do not carry out an automatic verification and vUML [13,17], which does, does not use the potential of temporal logic, performing a very reduced verification.

It should also be pointed out, though it has not been discussed here through lack of space, that the representative elements of both state and activity diagrams are included in this approach (except for synchronization states, events with parameters, and the dynamic creation and destruction of objects), something that cannot be said of other contributions in this field, in which few of the characteristics provided by UML (history states, deferred events, transitions fired by termination...) are dealt with.

Table 8-2. Related work

Tool	Formal Specification	Formal Verification	Language	UML
[12]	No automatic	No automatic	Promela (Spin)	Statechart
vUML [13,17]	Automatic No XMI	Automatic Limit to error states	Promela (Spin)	Statechart
HUGO [18,11]	Automatic XMI	Automatic Use Collaborations	Promela (Spin)	Statechart
veriUML [5,19,20]	Automatic XMI	No automatic	SMV	Statechart Class D.
TCM [8]	Automatic No XMI	No automatic	SMV	Activity D.
TABU [1,2,3]	Automatic XMI	Semi-automatic LTL	SMV	Statechart Activity D.

8.5 Conclusions and future lines of work

This paper presents a tool whose main aim is to integrate formal methods with semi-formal ones in such a way as to be transparent for the user. To be precise, it verifies the UML active behaviour using SMV. Although this is not a new idea, as far as we know at the present time, nowhere activity and state diagrams are jointly verified, using the former to represent the behaviour of the class operations.

However, the most innovative characteristic of the tool is that, in spite of using the potential of temporal logic to verify systems, the user need have no knowledge of its working. In addition, the user needs no knowledge of the structure of the specification obtained either, thus eliminating one of the major inconveniences of using formal methods.

As for future lines of work, some kind of treatment of the traces obtained in the verification when the property is not satisfied would seem to be of great interest. More precisely, that the representation of the traces should be visual instead of written, by using either some of the UML diagrams or an animated representation of the state and activity machines which could help the user to locate the error source very quickly.

8.6 References

[1]   M.E. Beato, M. Barrio-Solórzano, C. E. Cuesta and P. de la Fuente. UML Automatic Verification Tool with Formal Methods. Electronic Notes in Theoretical Computer Science, 127(4):3-16, 2005.

[2]   M. E. Beato, M. Barrio-Solórzano, C. E. Cuesta and P. de la Fuente. UML Automatic Verification Tool (TABU). Technical Report 04-09. Department of Computer Science, Iowa State University, pages 106-110. October 2004.

[3]   M. E. Beato. Verificación Formal del Comportamiento Activo de UML usando Métodos Formales. PhD thesis, Universidad de Valladolid, October 2004.

[4]   G. Booch, J. Rumbaugh and I. Jacobson. *The Unified Modeling Language*. Addison-Wesley, 1999

[5]   K. Compton, Y. Gurevich, J. Huggins and W. Shen.l An Automatic Verification Tool for UML, Technical Report CSE-TR-423-00, University of Michigan, 2000.

[6]   A. Darvas, I. Majzik and B. Benyó. Verification of UML Statechart Models of Embedded Systems. In B. Straube, E.J. Marinissen, Z. Kotasek, O.Novak, J. Hlavicka and R. Ruzicka, editors, *Proc. 5th IEEE Design and Diagnostics of Electronic Circuits and Systems Workshop (DDECS 2002)*, IEEE Computer Society TTTC, pages 70-77, April 2002.

[7]   M. B. Dwyer, G. S. Avrunin and J. C. Corbett. Patterns in Property Specifications for Finite-State Verification. In *Proceedings of the 21st International Conference on Software Engineering*, May 1999.

[8]   R. Eshuis. *Semantics and Verification of {UML} Activity Diagrams for Workflow Modelling*. PhD thesis, University of Twente, October 2002

[9]   T. Grose, G. Doney and S. Brodsky. *Mastering XMI. Java Programming with XMI, XML and UML*. OMG Press, 2002

[10]   D. Harel. STATECHARTS: A visual Formalism for Complex Systems. *Science of Computer Programming, North Holland*, 8:231-274, 1987.

[11]   A. Knapp and S. Merz. Model Checking and Code Generation for UML State Machines and Collaborations. In *Proc. 5th Wsh. Tools for System Design and Verification*, pages 59-64. Institut für Informatik, Universität Augsburg, Dominik Haneberg, Gerhard Schellhorn and Wolfgang Reeif, editors, 2002.

[12]   D. Latella, I. Majzik and M. Massink. Automatic verification of a behavioral subset of UML statechart diagrams using the SPIN model-checker. *Formal Aspects of Computing. The International Journal of Formal Methods*, 6(11):637-664, 1999.

[13]   J. Lilius and I. Porres. vUML: a Tool for Verifying UML Models. Technical Report TUCS 272, Turku Centre for Computer Science, Abo Akademi University, May 1999.

[14]   K. L. McMillan. *Symbolic Model Checking. An approach to the state explosion problem*. PhD thesis, Carnegie Mellon University, May 1992.

[15]   OMG. *Unified Modeling Language Specification v. 1.4*. OMG Document 01-09-67, 2001

[16]   OMG. *XML Metadata Interchange (XMI)*, OMG Document ad/98-10-05, 1998

[17]   I. Porres. *Modeling and Analyzing Software Behavior in UML*. PhD thesis, Department of Computer Science, Abo Akademi University, November 2001.

[18]   T. Schäfer, A. Knapp and S. Merz. Model Checking UML State Machines and Collaborations, *Proc. Wsh. Software Model Checking of Electronic Notes in Theoretical Computer Science*, 55(3):1-13, 2001.

[19]   W. Shen, K. Compton and J. Huggins. A validation Method for UML Model Based on Abstract State Machines. In R. Moreno-and A. Quesada-editors, *Proceeding of EUROCAST 2001*, pages 220-223, February 2001.

[20]   W. Shen, K. Compton and J. Huggins. A Toolset for Supporting UML Static and Dynamic Model Checking. In *COMPSAC*, pages 147-152. IEEE Computer Society, 2002.

Index